INSTRUCTOR'S MANUAL

AF870495

to accompany

PRINCIPLES OF ENGINEERING ECONOMY

SEVENTH EDITION

EUGENE L. GRANT

W. GRANT IRESON

RICHARD S. LEAVENWORTH

1807 175 YEARS OF PUBLISHING 1982

JOHN WILEY & SONS
New York Chichester Brisbane Toronto Singapore

ISBN 0-471-08439-5
Printed in the United States of America

10 9 8 7 6 5 4 3 2 1

CONTENTS

PREFACE

This instructor's Supplement contains the complete answers, and related calculations, for the problems in PRINCIPLES OF ENGINEERING ECONOMY, Seventh Edition. The amount of detail included in each solution varies from problem to problem but in general the new work in any chapter is given the most thorough treatment. Although most problems can be formulated and solved by two or more methods, only one method of solution is given here for each problem. It is important for the instructor to point out to students that different methods frequently offer advantages from a computational viewpoint, but that all methods render the same basic results, when correctly applied.

Comments and suggestions have been included in most chapters, and in some cases in individual problems, to assist instructors in making the most effective use of the problem material as a teaching aid.

Years of teaching engineering economy to a wide variety of students, from college freshmen to practicing engineers with many years of experience, has convinced the authors of the effectiveness of learning through problem solving. Many students who think that they understand the principles of engineering economy proceed to make serious errors when they attempt to apply these principles in solving large, unstructured problems. The instructor can help establish the significance of the principles and their applications by making the problems and their solutions an integral part of the course work.

Most of the problems in the text are intentionally simple and well structured, minimizing to some extent the time required by the student to solve them. The main objective is to emphasize some particular situation encountered in actual practice. Each problem also affords the instructor the opportunity to illustrate variations of these common situations and to demonstrate the different methods used to solve them.

In a great many of the problems presented, the numerical answer is ultimately translated into a yes-no, make-buy, repair-replace type of decision. The quantitative answers that lead to these decisions, however, are very important because the decision maker can use his or her own judgment regarding irreducibles and weigh the monetary advantages of the alternative courses of action against the non-quantifiable advantages and disadvantages. The decision maker is the responsible person. Therefore, as part of the overall training of engineers who are or will be decision makers, these problems can serve as the basis for broad discussions on the analysis of alternatives.

Many of the solutions in this supplement were done using a programmable electronic pocket calculator and a microcomputer. As such, they might be expected to be more accurate than those obtained by interpolation from the tables using a pocket or desk top calculator. This is particularly true for those problems for which the interest rate is the unknown variable. In using this supplement to check student solutions, emphasis should be put on problem formulation and analysis rather than on the precision of the final numerical answer.

The instructor may also wish to caution students using electronic pocket calculators that the accuracy of the tables in the text limits the number of significant figures which may justifiably be carried in any given calculation.

The authors would like to express their sincere appreciation to Mrs. JoAnne Christie Leavenworth for her care and patience in preparing most of the manuscript of this supplement.

Eugene L. Grant
W. Grant Ireson
Richard S. Leavenworth

November, 1981

No problems in Chapter 1.

CHAPTER 2

The Need for Criteria and Analytical Procedures in Making Decisions about Proposed Investments

General Notes:

Chapters 1 and 2 delineate eleven fundamental concepts of engineering economy that must be thoroughly understood by the practicing engineering economist. These concepts are briefly explained in these two chapters, but their significance becomes more clear as one studies the remaining chapters and the Appendixes. The new student to engineering economy should start thinking about problems in terms of these concepts, and the problems in Chapter 2 are designed to force him to begin analyzing problems in relationship to these concepts.

It is suggested that the teacher emphasize the logic of the methods in his discussion of the problems. Students can begin to formulate the problems for numerical solutions even though the numerical methods are first introduced in the next chapter (Chapter 3.) An incorrectly formulated problem cannot yield a correct solution even though the arithmetic may be perfect.

In non-numerical problems (such as those in Chapter 2) one can expect wide variety in the answers to the questions. The value and the validity of the proposed answers can be judged relative to the eleven concepts and the completeness of the answers. The following suggestions cover key points in the solutions, but they should not be considered the only points that might logically be considered.

2-1

Some prospective disbursements in connection with car ownership are for initial purchase, motor fuel, lubricants, tire repair and replacement (if necessary), repairs and maintenance, insurance, license fees and property taxes, parking fees and - possibly - parking tickets, traffic fines, and costs associated with accidents. Bus fare is the obvious disbursement in connection with the use of the public bus. Two important irreducibles are the time saved by car ownership (because of no waiting for buses and, presumably, faster transportation to and from the campus), and the disadvantage

2-1 Continued

of having no public transportation after 8 p.m. The major receipt is from the sale of the car at the end of the school year. Conceivably there might also be receipts from passengers who would agree to share expenses.

2-2

The disbursement to acquire the car has already been made; it cannot be changed by the choice between keeping the car for the school year and selling it at once. If the car can be sold for the amount that was paid for it, the analysis - in effect - is the same as before. If an immediate sale will bring less than the amount paid for the car, the analysis is the same as in Problem 2-1 except that the net resale price of the car takes the place of its purchase price.

2-3

(a) \$33, the amount obtainable if the ticket should be redeemed by the bus company.

(b) \$50, the amount you will have to pay to replace the ticket.

2-4

The ranch owner has two major alternatives; build a pole line to the nearest public utility power line or install and operate his own generation plant. There are several alternatives under each of these two main alternatives. The kind and sizes of the poles, the size of wire and the spacing of the poles will affect the initial cost, the line losses, and the maintenance expenditures. His own power generation plant might be fueled by one of several different fuels which in turn will influence the type of turbine, engine, or energy converter that he will need. Considerable differences will exist among the various alternative courses of action, and will be reflected in differences in the required initial investments, the operating costs, and the prospective lives of the installations.

The concepts that need to be discussed are:

1. Identification and definition of all alternatives.
2. Estimation of the consequences of each possible alternative.
3. Can all the differences in the consequences be reduced to monetary terms over time?
4. Separable decisions, so far as possible, should be made separately. What is the most economical system? Can it be financed?

2-4 Continued

5. Secondary criteria should be examined. How will variations in the needs in the future, and possible variations in fuel costs, affect the decision? How sensitive is the decision to the estimates?

2-5

Principal alternative courses of action for the boat owner to consider are:

A. Do nothing now. Continue to operate as in the past, having the engine repaired as necessary.

The effect of this course of action will be an increasing reduction of receipts due to the increasing rate of breakdowns and loss of goodwill will accelerate the reduction in income. From recent past experience he could estimate the number of days the boat will be "laid up" in the next year due to engine breakdowns. From that he could estimate the prospective loss of income and the prospective increase in repair expenditures.

B. Have engine completely overhauled now.

This requires an immediate outlay (or investment) of cash and the loss of some number of days of operation while the overhaul takes place. However, he could expect to lose fewer days of operation from engine breakdown than he would have without the overhaul. Goodwill should improve and the usual engine repair expenditures should be lower than for alternative A. He can convert the number of days of operation, having had the overhaul, into an estimate of the additional income he could expect to receive.

C. Purchase a rebuilt engine.

D. Purchase a new engine.

These actions would require larger initial investments than alternative B but should reduce the probability of lost time in the future and the probable repair expenses. Both the income and goodwill should be improved with a more reliable engine, and the owner should feel more comfortable about the safety of his passengers when he takes parties out fishing.

E. Buy a new boat and a new engine.

This alternative will require a much greater investment (cost of new boat and engine less the salvage value of the old boat and engine) but it should reduce the dangers of breakdown and the repair and maintenance expenditures. He should be able to operate more nearly at 100% of capacity, and the new boat might enable him to carry a larger party.

2-5 Continued

Irreducibles: Goodwill has been mentioned in the discussion of the several alternatives. Goodwill means more or fewer "repeat calls" for his boat from the same persons, and availability of a boat to serve the desires of the prospective customers is important to his business since he must depend upon repeat business. Good service, availability, and a feeling of safety on the boat are important to the prospective customers. His "reputation" is at stake, and he must consider more than just the cost of lost time in making his decision.

The owner assumes the responsibility for the safety of his passengers whenever he takes a party out fishing. An engine failure in rough water could result in many adverse effects on his future as a "party boat captain" and might result in his loss of license, cancellation of his liability insurance and even the loss of his boat. The very future of his business depends upon the reliability of his engine and boat.

A new engine (or a new boat and engine) will inspire confidence in his customers and will result in their selecting him over competing boat owners. He could use the fact that he has a new engine or a new boat and engine in his sales promotion campaign. It is very difficult to estimate the increased income that might be generated, so this advantage must be treated as an irreducible.

2-6

This is valuable as a class discussion problem in which the primary purpose is to identify "whose viewpoint" should be used in its solution. It is obvious that there are several different viewpoints that might be taken: the state, the city, the residents of the vicinity of the intersection, the general public, and the auto insurance companies. What principle should govern the decision? Is the city's viewpoint significantly different from that of the residents of the two streets? Who will pay, in the long run, for the consequences of the different alternatives? Why? How?

2-7

The alternative methods by which a college might provide janitorial service for its buildings are:

A. Provide its own service by employing:
   1. full-time unskilled labor from the local market
   2. part-time students and full-time supervisors
   3. a combination of these two labor groups.
B. Contract the work to a local janitorial contractor on a fixed price basis.

2-7 Continued

Some comments regarding the concepts and their application to this problem:

All alternatives need to be identified and carefully defined. The two primary and three subsidiary alternatives listed do not define all the possible subalternatives that might be identified. Alternatives may depend upon some degree of use of full-time and part-time personnel. Each variation will have some effects on the expenditures.

The expected consequences of each of the alternatives will include differences in the cost as well as in the time at which the functions will be performed. It is assumed that the quality of service has been defined and that the consequences of selecting any of the alternatives will not include a difference in the quality of the service.

The primary criterion will be the cost to the college to keep its buildings clean. A secondary criterion involves the uncertainty of the number of persons to be hired under each of the variations of Alternative A. The fact that alternative A-2 or A-3 may provide financial support for needy students might be an irreducible, unless the lack of that work would necessitate the allocation of additional scholarships.

The provision of janitorial service is a systems problem. There are many different kinds of activity and work that must be provided, and many different ways of providing each. Looking at the whole "system" and reduction of the work load to quantities of time by labor grade or classification will offer some opportunity for consideration of definitive combinations of men and machines (or mechanical or electrical aids.)

Whose viewpoint presumably is quite clear: the business manager for the college needs to minimize the total cost of providing a service to the college.

2-8

The determination of the acceptable alternative ways by which a married student can provide housing for his family while he is in college requires first that "acceptable housing" be defined. It is assumed here that he has children of school age, and that his wife must stay at home most of the time to care for pre-school children. Therefore, he must set some standards regarding proximity to public schools as well as to the college relative to public transportation or his own car, the number of bedrooms, and his ability to pay.

2-8 Continued

He has two major alternatives: rent or own. Each of these will also give several alternatives, but they can be summarized by three types of housing: single family units, apartments, mobile homes. Thus, he can either buy or rent a single family house, but he will probably have to choose from among many available houses. A similar situation may exist for apartments and mobile homes. In the case of rental, he may have another class of alternatives represented by rental from the college or rental from private owners.

His decision among the alternatives may be influenced by the effect of the location on his ability to earn supplemental income while going to school, or the availability of jobs for his wife that can be performed in the home.

Whose viewpoint obviously is the student's entire family. He can not make the decision without consideration of the effects of that decision on all the members of his family, as well as upon him.

His primary criterion is probably to minimize the total cost of housing for the duration of his educational program, with the understanding that the decision must be one that will contribute to the happiness of his family. The resources at his disposal (for down payment on a house, for example) and his available funds to meet monthly expenses will probably place an upper limit on his set of alternatives from which he must choose.

To determine the most economical housing, he must first examine all of the consequences of each choice without regard to the financing of the decision. After he has made an array of the alternatives with the equivalent annual cost or present worth of the future costs, he can then make the separable decision, "Can I meet the cash flow requirements for the most economical housing?" Separable decisions should be made separately.

He should tabulate, very carefully, all of the expenditures and incomes that might arise with each alternative, and also all of the differences that cannot be reduced to monetary terms, such as the fact that one alternative might provide him with more time to study because it does not involve maintenance (mowing the lawn, shoveling snow, etc.) These irreducibles must be considered in the final decision process and their advantages or disadvantages must be weighted against the monetary differences that exist among the feasible alternatives.

CHAPTER 3

Equivalence

3-1

	End of Year	Interest Due(9%)	Total Before Year-End Payment	Year-End Payment	Owed After Year-End Payment
	0				$10,000
	1	$900	$10,900	$2,900	8,000
Plan II	2	720	8,720	2,720	6,000
	3	540	6,540	2,540	4,000
	4	360	4,360	2,360	2,000
	5	180	2,180	2,180	0
	0				$10,000.00
	1	$900	$10,900	$2,570.90	8,329.10
Plan III	2	749.62	9,078.72	2,570.90	6,507.82
	3	585.70	7,093.52	2,570.90	4,522.62
	4	407.04	4,929.66	2,570.90	2,358.76
	5	212.29	2,571.05	2,571.05*	0.00

*difference due to rounding error of (A/P) factor.

Total repayment	5 yrs.	10 yrs.
Plan II	$12,700	$14,950
Plan III	12,854.65	15,580

3-2

(Headings same as Problem 3-1 and Table 3-1)

0				$4,000
1	$440	$ 4,440	$1,240	3,200
2	352	3,552	1,152	2,400
3	264	2,664	1,064	1,600
4	176	1,776	976	800
5	88	888	888	0

3-3

(Headings same as Problem 3-1 and Table 3-1.)

0				$4,000.00
1	$440.00	$4,440.00	$1,082.28	3,357.72
2	369.35	3,727.07	1,082.28	2,644.79
3	290.93	2,935.72	1,082.28	1,853.44
4	203.86	2,057.30	1,082.28	975.02
5	107.25	1,082.27	1,082.27	0.00

3-4

(Headings same as Problem 3-1 and Table 3-1.)

	0				$2,000
Plan I	1	$240	$2,240	$240	2,000
	2	240	2,240	240	2,000
	3	240	2,240	240	2,000
	4	240	2,240	2,240	0
	0				$2,000
Plan II	1	$240	$2,240	$740	1,500
	2	180	1,680	680	1,000
	3	120	1,120	620	500
	4	60	560	560	0
	0				$2,000.00
Plan III	1	$240	$2,240.00	658.46	1,581.54
	2	189.79	1,771.33	658.46	1,112.87
	3	133.54	1,246.41	658.46	587.95
	4	70.55	658.50	658.50	0.00
	0				$2,000
Plan IV	1	$240.00	$2,240.00	$ 0.00	2,240.00
	2	268.80	2,508.80	0.00	2,508.80
	3	301.06	2,809.86	0.00	2,809.86
	4	337.18	3,147.04	3,147.04	0.00

3-5

(Headings same as Problem 3-1 and Table 3-1.)

	0				$5,000
	1	$450	$5,450	$1,450	4,000
	2	360	4,360	1,360	3,000
9%	3	270	3,270	1,270	2,000
	4	180	2,180	1,180	1,000
	5	90	1,090	1,090	0

3-5 continued

	0				$5,000
	1	$750	$5,750	$1,750	4,000
	2	600	4,600	1,600	3,000
15%	3	450	3,450	1,450	2,000
	4	300	2,300	1,300	1,000
	5	150	1,150	1,150	0

Total payments at 9% ($6,350) are $900 less than those at 15%($7,250.)

3-6

(Headings same as Problem 3-1 and Table 3-1.)

	0				$5,000
	1	$600	$5,600	$1,850	3,750
4 yrs.	2	450	4,200	1,700	2,500
	3	300	2,800	1,550	1,250
	4	150	1,400	1,400	0

	0				$5,000
	1	$600	$5,600	$1,100	4,500
	2	540	5,040	1,040	4,000
	3	480	4,480	980	3,500
	4	420	3,920	920	3,000
10 yrs	5	360	3,360	860	2,500
	6	300	2,800	800	2,000
	7	240	2,240	740	1,500
	8	180	1,680	680	1,000
	9	120	1,120	620	500
	10	60	560	560	0

The 10-year payment series is $1,800 greater ($8,300) than the 4-year series ($6,500).

3-7

(Headings same as Problem 3-1 and Table 3-1.)

	0				$2,000
	1	$300	$2,300	$300	2,000
Plan	2	300	2,300	300	2,000
I	3	300	2,300	300	2,000
	4	300	2,300	2,300	0

3-7

continued (Headings same as Problem 3-1 and Table 3-1.)

	0				$2,000
	1	$300	$2,300	$800	1,500
Plan	2	225	1,725	725	1,000
II	3	150	1,150	650	500
	4	75	575	575	0
	0				$2,000.00
	1	$300	$2,300.00	$700.54	1,599.46
Plan	2	239.92	1,839.38	700.54	1,138.84
III	3	170.83	1,309.67	700.54	609.13
	4	91.37	700.50	700.50	0.00
	0				$2,000.00
	1	$300.00	$2,300.00	$ 0.00	2,300.00
Plan	2	345.00	2,645.00	0.00	2,645.00
IV	3	396.75	3,041.75	0.00	3,041.75
	4	456.26	3,498.01	3,498.01	0.00

3-8

(Headings same as Problem 3-1 and Table 3-1.)

	0				$10,000
	1	$600	$10,600	$1,600	9,000
	2	540	9,540	1,540	8,000
	3	480	8,480	1,480	7,000
	4	420	7,420	1,420	6,000
Plan	5	360	6,360	1,360	5,000
II	6	300	5,300	1,300	4,000
	7	240	4,240	1,240	3,000
	8	180	3,180	1,180	2,000
	9	120	2,120	1,120	1,000
	10	60	1,060	1,060	0
	0				$10,000.00
	1	$600.00	$10,600.00	$1,358.68	9,241.32
	2	554.48	9,795.80	1,358.68	8,437.12
	3	506.23	8,943.35	1,358.68	7,584.67
	4	455.08	8,039.75	1,358.68	6,681.07
Plan	5	400.86	7,081.93	1,358.68	5,723.25
III	6	343.40	6,066.65	1,358.68	4,707.98
	7	282.48	4,990.45	1,358.68	3,631.77
	8	217.91	3,849.68	1,358.68	2,491.00
	9	149.46	2,640.46	1,358.68	1,281.78
	10	76.90	1,358.68	1,358.68	0.00

CHAPTER 4

Interest Formulas

4-1

Semiannually: $i_e = (1+0.12/2)^2 - 1 = (1.06)^2 - 1 = 0.1236 = \underline{12.36\%}$

Quarterly: $i_e = (1+0.12/4)^4 - 1 = (1.03)^4 - 1 = 0.1255 = \underline{12.55\%}$

Monthly: $i_e = (1+0.12/12)^{12} - 1 = (1.01)^{12} - 1 = 0.1268 = \underline{12.68\%}$

4-2

Nominal = 12(1.5%) = $\underline{18\%}$.

Effective = $(1.015)^{12} - 1 = 0.1956 = \underline{19.56\%}$.

4-3

$i_e = (1+0.15/12)^{12} - 1 = (1.0125)^{12} - 1 = 0.16075 = \underline{16.075\%}$.

By logarithms: 12 log 1.0125 = 12(0.005395)

= 0.06474; $\log^{-1}$ 0.06474 = 1.16075

$i_e = 1.16075 - 1 = 0.16075 = \underline{16.075\%}$.

4-4

Daily: $i_e = (1+0.15/365)^{365} - 1 = 1.16180 - 1 = 0.16180$

= $\underline{16.180\%}$. By logarithms:

365 log (1+0.15/365) = 365(0.0001784) = 0.06513

$i_e = \log^{-1} 0.06513 - 1 = 0.16180 = \underline{16.180\%}$.

Continuous: $i_e = e^r - 1 = e^{0.15} - 1 = 1.16183 - 1 = \underline{0.16183}$

= $\underline{16.183\%}$.

4-5

A formula to relate a series of beginning of period deposits, B, in a sinking fund for n periods with interest at i per period to the amount accumulated, F, at the end of the nth period, may be derived directly, or the end-of-period formulas given in Chapter 4 may be used to derive the formulas.

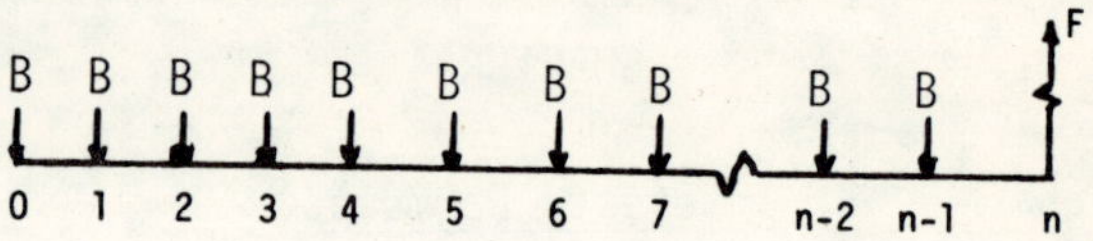

F can be expressed as the summation of a series:

$$F = B[(1+i)+(1+i)^2 + ----- + (1+i)^{n-1} + (1+i)^n]$$

Multiplying both sides of the equation by (1+i), we have

$$F(1+i) = B[(1+i)^2 + (1+i)^3 + ----- + (1+i)^n + (1+i)^{n+1}]$$

Subtracting the second equation from the first, we obtain

$$F - F(1+i) = B[(1+i) - (1+i)^{n+1}]$$

That equation can be cleared up to

$$F = B\,\frac{[(1+i) - (1+i)^{n+1}]}{1 - (1+i)} = B\,\frac{(1+i)[1-(1+i)^n]}{1 - (1+i)} \quad \text{and then}$$

$$B = F\left[\frac{1 - (1+i)}{[1-(1+i)^n](1+i)}\right] = F\left[\frac{i}{[(1+i)^n - 1](1+i)}\right]$$

4-6

Let B' be the end-of-period deposit in a sinking fund that will accumulate to F, at interest = i, in n periods. The sinking fund factor multiplied by F will give B':

$$B' = \frac{i}{(1+i)^n - 1}F$$

; At each period the end-of-period amount B' is equal to the beginning-of-period amount B(F/P,i%,n), or to express B in terms of B', B=B'(P/F,i%,1)

Substituting the formula for (P/F,i%,n) in the equation B', we have

$$B = \left[\frac{i}{(1+i)^n - 1} \times \frac{1}{(1+i)}\right]F = (A/F,i\%,n)(P/F,i\%,1)\,F$$

4-6 continued

From the formula in 4-5:

$$B = \left[\frac{0.15}{[(1+0.15)^{10}-1](1+0.15)}\right] F = \underline{0.042828\ F}.$$

From Table D-20: B = 0.04925(0.8696) F = $\underline{0.042828F}$.

4-7

(A/G,i%,n) = G/i -[nG/i](A/F,i%,n)

= G[(1/0.11) - (10/0.11)(0.05980)] = $\underline{G(3.6545)}$.

By the formula:

$$A = G\ [(1/0.11) - 10/\{(1+0.11)^{10} - 1\}]$$

$$= G\ [9.0909 - 5.4365] = \underline{G(3.6544)}.$$

4-8

(F/A,16%,10) = 21.321 from Table D-21.

(a) $(F/A,16\%,10) = 1 + \sum_{t=1}^{9} (F/P,16\%,t)$

= 1+1.16+1.3456+1.5609+1.8106+2.1003+2.4364

+ 2.8262+3.2784+3.803 = $\underline{21.3214}$.

(b) (F/A,i%,n) = 1/(A/F,i%,n)

= 1/(A/F,16%,10) = 1/0.04690 = $\underline{21.322}$.

4-9

(A/P,i%,n); (A/P,9%,6) = 0.22292 from Table D-14.

(a) $(A/P,i\%,n) = [\sum_{t=1}^{n} (P/F,i\%,t)]^{-1}$.

$(A/P,9\%,6) = [0.9174+0.8417+0.7722+0.7084+0.6499 + 0.5963]^{-1} = [4.4859]^{-1} = \underline{0.22292}$.

(b) (A/P,i%,n) = 1/(P/A,i%,n).

(A/P,9%,6) = 1/(4.486) = $\underline{0.22292}$.

4-10

$i = \sqrt[18]{F/P} - 1.$

$(\log F/P)/18 = \log (2{,}689.37/1{,}000)/18$

$= 0.42965/18 = 0.023869$;

$i = \log^{-1}(0.023869) - 1 = 0.0565 = \underline{5.65\%}.$

4-11

$(1+i) = e^r.$ Thus $r = \ln(1+i)$

$= \ln(1 + 0.12) = 0.113329 = \underline{11.3329\%}.$

4-12

$(1+i_e) = (1+r/365)^{365}$

$365 \log(1+r/365) = \log(1+0.12) = 0.049218.$

$\log^{-1}(1+r/365) = \log^{-1}(0.049218/365) = 1.000310538$

$= 1+r/365$ from which $r = \underline{11.3346\%}.$

This solution differs from that in Problem 4-11 by only 0.000017, hardly a significant amount.

4-13

$(A/P,i\%,n) = [i(1+i)^n]/[(1+i)^n-1]$

$\underline{n=5}$: $(A/P,22\%,5) = [0.22(1.22)^5]/[1.22^5 - 1]$

$= 0.5946/1.7027 = \underline{0.34921}.$

From Appendix D:

$(A/P,20\%,5) = 0.33438$; $(A/P,25\%,5) = 0.37185.$

By linear interpolation for 22%,

$(A/P,22\%,5) = 0.33438 + (2/5)(0.37185 - 0.33438)$

$= 0.34939.$ This differs from the correct value by +0.052%.

4-13 continued

n=10: $(A/P,22\%,10) = [0.22(1.22)^{10}]/[1.22^{10} - 1]$

$= 1.60702/6.30463 = \underline{0.25489}.$

From Appendix D:

$(A/P,20\%,10) = 0.23852$; $(A/P,25\%,10) = 0.28007$.

By linear interpolation for 22%,

$(A/P,22\%,10) = 0.23852 + (2/5)(0.28007-0.23852)$

$= 0.25514$. This differs from the correct value by +0.098%.

n=20: $(A/P,22\%,20) = [0.22(1.22)^{20}]/[1.22^{20} - 1]$

$= 11.73868/52.35764 = \underline{0.22420}.$

From Appendix D:

$(A/P,20\%,20) = 0.20536$; $(A/P,25\%,20) = 0.25292$.

By linear interpolation for 22%,

$(A/P,22\%,10) = 0.20536 + (2/5)(0.25292 - 0.20536)$
$= 0.22438$. This differs from the correct value by +0.080%. In all cases the interpolated value exceeds the actual by a very small amount.

4-14

$(P/A,i\%,n) = [(1+i)^n - 1]/[i(1+i)^n]$

n=5: $(P/A,22\%,5) = [(1.22)^5 - 1]/[0.22(1.22)^5]$

$= 1.7027/0.5946 = \underline{2.8636}.$

From Appendix D:

$(P/A,20\%,5) = 2.991$; $(P/A,25\%,5) = 2.689$.

<u>4-14</u> continued

By linear interpolation for 22%,

$(P/A,22\%,5) = 2.991 - (2/5)(2.991-2.689)$

$= 2.8702$. This differs from the correct value by +0.230%.

<u>n=10</u>: $(P/A,22\%,10) = [(1.22)^{10} - 1]/[0.22(1.22)^{10}]$

$= 6.3046/1.6070 = \underline{3.9232}$.

From Appendix D:

$(P/A,20\%,10) = 4.192$; $(P/A,25\%,10) = 3.571$.

By linear interpolation for 22%,

$(P/A,22\%,10) = 4.192 -(2/5)(4.192-3.571)$

$= 3.9436$. This differs from the correct value by +0.520%.

<u>n=20</u>: $(P/A,22\%,20) = [(1.22)^{20} - 1]/[0.22(1.22)^{20}]$

$= 52.3576/11.7387 = \underline{4.4603}$.

From Appendix D:

$(P/A,20\%,20) = 4.870$; $(P/A,25\%,20) = 3.954$.

By linear interpolation for 22%,

$(P/A,22\%,20) = 4.870 -(2/5)(4.870 - 3.954)$

$= 4.5063$. This differs from the correct value by +0.971%. In all cases, the interpolated value exceeds the actual value. This difference, while increasing, is less than 1%.

4-15

For n in the range compared (5-20), both factors give higher than actual values when interpolation is used. The (A/P) factor gives closer results (with an error of less than 0.1%) than the (P/A) factor (with an error of less than 1%, an order of magnitude greater).

4-16

n=5: (P/A,2%,5) = 4.713 ; (P/A,10%,5) = 3.791.

By linear interpolation for 6%,

$$(P/A,6\%,5) = 4.713 - (4/8)(4.713-3.791) = 4.252$$

From Table D-11, (P/A,6%,5) = 4.212. The interpolated value differs from the actual by +0.950%.

n=10: (P/A,2%,10) = 8.983 ; (P/A,10%,10) = 6.144

By interpolation for 6%,

$$(P/A,6\%,10) = 8.983 - (1/2)(8.983 - 6.144) = 7.564.$$

From Table D-11, (P/A,6%,10) = 7.360. The interpolated value differs the actual by +2.77%.

n=20: (P/A,2%,20) = 16.351 ; (P/A,10%,20) = 8.514.

By interpolation for 6%,

$$(P/A,6\%,20) = 16.351 - (1/2)(16.351 - 8.514) = 12.433.$$

From Table D-11, (P/A,6%,20) = 11.470. The interpolated value differs from the actual by +8.40%. The percentage error increases rapidly as n increases. (Compared to Problems 4-13 through 15, smaller values of i lead to greater error.)

CHAPTER 5

Solving Interest Problems

5-1

$P = \$500(P/A,5\%,20) + \$100(P/G,5\%,20)$

$= \$500(12.462) + \$100\ (98.488) = \underline{\$16,080}$

5-2

$P = \$2,400(P/A,5\%,20) - \$100(P/G,5\%,20)$

$= \$2,400(12.462) - \$100(98.488) = \underline{\$20,060}$

5-3

The semiannual payment is \$100,000(A/P,3.5%,60)

=\$100,000(0.04009) = \$4,009

40 payments remain after 20 have been made, thus the principal remaining is $P_{20} = \$4,009(P/A,3.5\%,40) = \$4,009(21.355)$

= \$85,612. \$100,000 - \$85,612 = \$14,388 of the original principal has been retired.

5-4

$P_{7192} = \$1,000(P/A,5.5\%,10) = \$1,000(7.538) = \$7,538.$

$A = \$7,538(A/F,5.5\%,18) = \$7,538(0.03392) = \underline{\$255.70}$

5-5

$A = [\$10,000 + \$5,000(P/F,10\%,5)](A/P,10\%,10)$

$+ \$1,000 + \$500(A/G,10\%,10)$

$= [\$10,000 + \$5,000(0.6209)](0.16275) + \$1,000$

$+ \$500(3.725) = \underline{\$4,995.26}.$

5-6

$(A/P,i\%,15) = \$51/\$600 = 0.0850$

$(A/P,3\%,15) = 0.08377$; $(A/P,3.5\%,15) = 0.08683$

by interpolation, i = 3.2% per month. The nominal rate per year is 12(3.20) = 38.4%. The effective interest rate per year is $(1.032)^{12} - 1 = 46.0\%$.

5-7

(a) $(A/P,i\%,10) = \$1,300/\$8,000 = 0.1625$

$(A/P,9\%,10) = 0.15582$; $(A/P,10\%,10) = 0.16275$

by interpolation, $i = 9.96\% \simeq 10\%$.

(b) $(A/P,i\%,20) = \$680/\$6,000 = 0.11333$

$(A/P,9\%,20) = 0.10955$; $(A/P,10\%,20) = 0.11746$

by interpolation, i = 9.5%.

(c) $(A/P,i\%,15) = \$400/\$6,000 = 0.06667 = 1/n$

Therefore i = 0%.

(d) $A = 0 = -\$10,000(A/P,i\%,20)+\$300+\$50(A/G,i\%,20)$

Try 3.5%; $-\$10,000(0.07036)+\$300+\$50(8.365) = +\$14.65$

Try 4.0%; $-\$10,000(0.07358)+\$300+\$50(8.209) = -\$25.35$

by interpolation, i = 3.7%.

(e) Since the capital recovery factor (A/P) approaches i as n approaches infinity, A = Pi from which i = \$225/\$5,000 = 0.045, or i = 4.5%.

5-8

The differences between the two plans, purchase vs. lease, are the purchase and resale values vs. the beginning-of-year lease payments. The interest rate that will make the present (for example) of these two plans equal is the prospective rate of return on the investment in the land purchase. Thus equate:

5-8 continued

$-\$80,000 + \$100,000(P/F,i\%,15) = -\$5,000[1 + (P/A,i\%,14)]$

at i = 7%; $-\$75,000 + \$100,000(0.3624)+\$5,000(8.245) = \$4,965$

at i = 8%; $-\$75,000 + \$100,000(0.3152)+\$5,000(8.244) = -\$2,260$

by interpolation, i = 7.7%.

5-9

This problem can be visualized as substituting \$55,000 - \$5,000 now for \$5,000 at the end of each of the next 14 years. Thus $(A/P,i\%,14) = \$5,000/\$50,000 = 0.10000$. By interpolation, i = 4.8%. It may be interpreted as the interest rate that would be paid on a loan of \$50,000 if the decision is made not to prepay the rent.

5-10

$(A/P,i\%,16) = \$14.44/\$200 = 0.07220$

$(A/P,1.5\%,16) = 0.07077$; $(A/P,2\%,16) = 0.07365$;

by interpolation, i = 1.75% per week. This is a nominal 90.91% per year and an effective $(1+0.01748)^{52} - 1 = 146.3\%$ per year.

5-11

The purchaser must be able to sell the land 10 years hence for the compound amount of all expenditures up to that time. Thus the selling price must be:

$$F = \$20,000(F/P,15\%,10) + \$400(F/A,15\%,10)$$

$$+ \$40(A/G,15\%,10)(F/A,15\%,10)$$

$$= \$20,000(4.0456) + \$400(20.304) + \$40(3.383)(20.304)$$

$$= \$80,912 + \$8,122 + \$2,747 = \underline{\$91,781}$$

5-12

Since some of the payments are intermittent, they must be converted to an equivalent P(or F) before conversion to a uniform A for 30 years all at 9% interest.

5-12 continued

$P = \$10,000[1 + (P/F,9\%,10) + (P/F,9\%,20)]$

$= \$10,000(1 + 0.4224 + 0.1784) = \$16,008$

$A = \$16,008(A/P,9\%,30) + 2,000 = \$16,008(0.09734)$

$+ \$2,000 = \underline{\$3,558.22}.$

5-13

(a) $F = \$5,000(F/P,1.5\%,80) = \$5,000(3.2907) = \underline{\$16,454}.$

(b) $F = \$5,000(F/P,6\%,20) = \$5,000(3.2071)$

$= \$16,036$; Difference is $\underline{\$418}$.

5-14

$F_{10} = \$5,000(F/P,1.5\%,40) = \$5,000(1.8140)$

$= \$9,070$; $F_{20} = (\$9,070 - \$5,000)(F/P,1.5\%,40)$

$= \$4,070(1.8140) = \underline{\$7,383}.$

5-15

(a) $(A/P,i\%,20) = \$600/\$5,000 = 0.12000$

$(A/P,10\%,20) = 0.11746$; $(A/P,11\%,20) = 0.12558$;

by interpolation; $i = \underline{10.3\%}$.

(b) $(A/P,i\%,8) = \$500/\$3,000 = 0.16667$

$(A/P,6\%,8) = 0.16104$; $(A/P,7\%,8) = 0.16747$;

by interpolation, $i = \underline{6.9\%}$.

(c) $\$50,000(A/P,i\%,15) = \$8,500 + \$15,000(A/F,i\%,15)$

@16%; $-\$50,000(0.17936)+\$8,500+\$15,000(0.01936) = -\$177.6$

@15%; $-\$50,000(0.17102)+\$8,500+\$15,000(0.02102) = +\$264.3$

by interpolation, $i = 15.6\%$

5-16

$\$40(1+i) = \$50$; $i = (\$50/\$40) - 1 =$ 25% per week.

$52(25\%) = 1{,}300\%$ per year nominal;

$(1.25)^{52} - 1 = 10{,}947{,}544\%$ per year effective.

5-17

(a) $\$4{,}000(P/F,11\%,12) = \$4{,}000(0.2858) = \$1{,}143.20$

(b) $\$6{,}000(A/P,11\%,20) = \$6{,}000(0.12558) = \$753.48$

(c) $\$50{,}000(P/F,11\%,40) = \$50{,}000(0.0154) = \$770$

(d) $A = Pi = \$10{,}000(0.11) = \$1{,}100$

5-18

20 years: $\$50{,}000(A/P,9\%,20) = \$50{,}000(0.10955) = \$5{,}477.5$.

30 years: $\$50{,}000(A/P,9\%,30) = \$50{,}000(0.09734) = \$4{,}867$.

The difference is only $610.5 per year.

5-19

20 - yr mortgage: $\$5{,}477.5(P/A,9\%,5) =$

$\$5{,}477.5(3.890) = \$21{,}307$.

30 - yr mortgage: $\$4{,}867(P/A,9\%,15) = \$4{,}867(8.061) = \$39{,}233$.

5-20

$F = \$24(F/P,6\%,359) = \$24(1.06)^{359} = \$29{,}175{,}403{,}540$. This sum has been said to approximate the value of Manhattan today.

5-21

$$P = \$1{,}000(P/F,13\%,5) + \$3{,}000(P/F,13\%,10)$$

$$+ \$5{,}000[(P/F,13\%,15) + (P/F,13\%,20)]$$

$$= \$1{,}000(0.5428) + \$3{,}000(0.2946)$$

$$+ \$5{,}000(0.1599 + 0.0868) = \$2{,}660.1$$

$$A = \$2{,}660.1(A/P,13\%,25) = \$2{,}660.1(0.13643) = \$362.92$$

5-22

$F = \$2,000(F/P,1.5\%,12) + \$200(F/A,1.5\%,12)$

$+ \$20(A/G,1.5\%,12)(F/A,1.5\%,12)$

$= \$2,000(1.1956) + \$200(13.041)+\$20(5.323)(13.041) = \underline{\$5,995.35}.$

$A_{12} = 11(\$20) + \$200 = \underline{\$420}.$

5-23

$\$750 = \$50(P/A,i\%,8) + \$20(P/G,i\%,8) + \$1,500(P/F,i\%,8)$

@18%; $-\$750 + \$50(4.078) + \$20(12.633) + \$1,500(0.2660) = +\$105.56$

@20%; $-\$750 + \$50(3.837) + \$20(9.883) + \$1,500(0.2326) = -\$11.59$

by interpolation, $i = \underline{19.8\%}$.

5-24

$(A/P,i\%,24) = \$84/\$1,500 = 0.0560$

$(A/P,2.5\%,24) = 0.05591$; $(A/P,3\%,24) = 0.05905$;

by interpolation, i = 2.51% per month. This is 12(2.51) = 24.2% nominal per year and $(1.0251)^{12} - 1 = 27.04\%$ effective per year.

5-25

$(A/P,i\%,24) = \$304.07/\$6,000 = 0.05068$

$(A/P,1.5\%,24) = 0.04992$; $(A/P,2\%,24) = 0.05287$;

by interpolation, i = 1.628% per month, which is nominally 12(1.628) = 19.54% and effectively $(1.01628)^{12}-1 = 21.39\%$ per year. The insurance should be considered part of the financing because it is required by the lender to protect the lender.

5-26

$P_{65} = \$2,500(P/A,2\%,100) = \$2,500(43.098) = \$107,745$

(a) $A_{20} = \$107,745(A/F,2\%,160) = \$107,745(0.000878) = \underline{\$94.64}$

(b) $A_{40} = \$107,745(A/F,2\%,100) = \$107,745(0.00320) = \underline{\$344.78}$

5-27

P_{65} = \$2,500/0.02 = \$125,000

(a) A_{20} = \$125,000(A/F,2%,160) = \$125,000(0.000878) = <u>\$109.79</u>

(b) A_{40} = \$125,000(A/F,2%,100) = \$125,000(0.00320) = <u>\$400.00</u>

The deposits would increase by 16% in each case.

5-28

Since the deposits after the first are to be made every 4th interest period, they can be converted to equivalent uniform period amounts by \$1,000(A/F,1.5%,4) = \$1,000(0.24444) = \$244.44

Then, F = \$1,000(F/P,1.5%,80) + \$244.44(F/A,1.5%,80)

= \$1,000(3.2907) + \$244.44(152.711) = <u>\$40,619</u>.

CHAPTER 6

Equivalent Uniform Annual Cash Flow

6-1

	Plan I	Plan II
Interest on land investment		
\$75,000(0.09)	\$6,750	
100,000(0.09)		\$9,000
CR on structures		
\$75,000(A/P,9%,30) = \$75,000(0.09734) =	7,301	
\$150,000(A/P,9%,50)- \$30,000(A/F,9%,50)		
= \$150,000(0.09123)-\$30,000(0.00123) =		13,648
Equivalent uniform annual disbursements		
\$7,000+\$3,000(P/A,9%,10)0.09		
=\$7,000+\$3,000(6.418)0.09 =	8,733	
\$4,000		4,000
	\$22,784	\$26,648

6-2

	Type A	Type B
CR for Pipe		
\$120,000(A/P,6%,60)=\$120,000(0.06188)	\$7,426	
\$80,000(A/P,6%,30)=\$80,000(0.07265)		\$5,812
CR for Pumps		
\$15,000(A/P,6%,20)=\$15,000(0.08718)	1,308	
\$20,000(A/P,6%,20)=\$20,000(0.08718)		1,744
Equivalent Uniform Annual Energy Cost		
\$3,000+\$60(A/G,6%,60)		
= \$3,000 + \$60(14.791)	3,887	
\$4,000 + \$80(14.791)		5,183
	\$12,621	\$12,739

6-3

	Concrete
CR = \$350,000(A/P,7%,90) = \$350,000(0.07016) =	\$24,556
Annual Upkeep	2,500
	\$26,556

6-3 continued

	Wood
CR:	
Earth fill: $50,000(A/P,7%,90)=$50,000(0.07016)=	$3,508
Bleachers: $100,000(A/P,7%,30)=$100,000(0.08059) =	8,059
Seats: $40,000(A/P,7%,15)=$40,000(0.10979) =	4,392
Painting: $10,000(A/P,7%,3)=$10,000(0.38105) =	3,810
	$19,769

6-4

CR:	Type Y	Type Z
$8,400(A/P,16%,6) = $8,400(0.27139) =	$2,280	
$10,800(A/P,16%,6) = $10,800(0.21708)=		$2,344
Annual operating costs	1,700	1,500
Extra annual income taxes		200
	$3,980	$4,044

6-5

CR:	4-Years	5-Years
$8,400(A/P,10%,4)-$2,400(A/F,10%,4)		
=$8,400(0.31547)-$2,400(0.21547)=	$2,133	
$8,400(A/P,10%,5)-$1,500(A/F,10%,5)		
=$8,400(0.26380)-$1,500(0.16380) =		$1,970
Operation and Maintenance:		
$2,400+$400(A/G,10%,4)=$2,400+$400(1.381) =	2,952	
$2,400+$400(A/G,10%,5)=$2,400+$400(1.810) =		3,124
	$5,085	$5,094

6-6

CR:	Mach. J	Mach. K
$50,000(A/P,12%,12)-$14,000(A/F,12%,12)		
=$50,000(0.16144)-$14,000(0.04144) =	$7,492	
$30,000(A/P,12%,12)=$30,000(0.16144) =		$4,843
Operation & Maintenance:		
$6,000+$300(A/G,12%,12)=$6,000+$300(4.190)=	7,257	
$750+$100(4.190) =	1,169	
$8,000+$500(4.190) =		10,095
	$15,918	$14,938

6-7

	Concrete	Frame
CR:		
$116,000(A/P,9%,50)=$116,000(0.09123) =	$10,583	
$60,000(A/P,9%,25)=$60,000(0.10181) =		$6,109
Operation and Maintenance:	1,000	1,800
Insurance: $1.50[$400,000+(0.75)$116,000]/$1,000=	730	
$4[$400,000+(0.75)$60,000]/$1,000 =		1,780
Property Taxes: 0.015($116,000) =	1,740	
0.015($60,000) =		900
Extra annual income tax	545	
	$14,598	$10,589

6-8

CR: $116,000(A/P,18%,50)=$116,000(0.18005)=	$20,886	
$60,000(A/P,18%,25)=$60,000(0.18292) =		$10,975
Operation and Maintenance:	1,000	1,800
Insurance: (See 6-7)	730	1,780
Property Taxes	1,740	900
	$24,356	$15,455

6-9

	Model H	Model K
CR:		
$9,000(A/P,20%,3)-$2,500(A/F,20%,3)		
=$9,000(0.47473)-$2,500(0.27473)		
=$4,273 - $687 =	$3,586	
$11,000(A/P,20%,4)-$1,500(A/F,20%,4)		
=$11,000(0.38629)-$1,500(0.18629)		
=$4,249 -$279 =		$3,970
Operating & Maintenance costs		
$2,200+$300(A/G,20%,3)		
=$2,200+$300(0.879) =	2,464	
$1,700+$300(A/G,20%,4)		
=$1,700+$300(1.274)		2,082
	$6,050	$6,052

There is no real economic difference between the two models. Since the drive have a strong preference for Model H, probably it should be chosen.

6-10

CR:	Unit A	Unit B
$40,000(A/P,14%,8)=$40,000(0.21557) =	$8,623	
$50,000(A/P,14%,10)-$5,000(A/F,14%,10)		
=$50,000(0.19171)-$5,000(0.05171) =		$9,327
Annual disbursements	1,800	1,200
Add'l income tax		550
	$10,423	$11,077

6-11

CR:	Unit A	Unit B
$40,000(A/P,25%,8)=$40,000(0.30040)	$12,016	
$50,000(A/P,25%,10)-$5,000(A/F,25%,10)		
=$50,000(0.28007)-$5,000(0.03007)		$13,853
Annual disbursements	1,800	1,200
	$13,816	$15,053

6-12

CR:	System E	System F
$22,000(A/P,15%,6)-$1,000(A/F,15%,6)		
=$22,000(0.26424) - $1,000(0.11424) =	$5,699	
$36,000(A/P,15%,10)-$3,000(A/F,15%,10)		
=$36,000(0.19925)-$3,000(0.04925) =		$7,025
Annual operating and maintenance cost	8,400	4,900
Extra annual income tax		1,850
	$14,099	$13,775

6-13

CR:	System E	System F
$22,000(A/P,30%,6)-$1,000(A/F,30%,6)		
=$22,000(0.37839)-$1,000(0.07839) =	$8,246	
$36,000(A/P,30%,10)-$3,000(A/F,30%,10)		
=$36,000(0.32346)-$3,000(0.02346) =		$11,574
Annual operating and maintenance cost	8,400	4,900
	$16,646	$16,474

6-14

CR:	Unit X	Unit Y
$16,000(A/P,20%,10)-$2,000(A/F,20%,10)		
=$16,000(0.23852)-$2,000(0.03852) =	$3,739	
$24,000(0.23852)-$3,000(0.03852) =		$5,609
Annual costs:	9,000	7,000
$30(A/G,20%,10) = $30(3.074) =	92	
Inc. tax = $650+$15(3.074) =		696
	$12,831	$13,305

6-15

$$P(A/P,i\%,n) - S(A/F,i\%,n)$$

$$= P\left\{\frac{i(1+i)^n}{(1+i)^n-1}\right\} - S\left\{\frac{i}{(1+i)^n-1}\right\}$$

(a) $(P-S)(A/P,i\%,n) + Si$

$$= P\left\{\frac{i(1+i)^n}{(1+i)^n-1}\right\} - S\left\{\frac{i(1+i)^n}{(1+i)^n-1}\right\} + Si$$

$$Si\left\{1 - \frac{(1+i)^n}{(1+i)^n-1}\right\}$$

$$= Si\left\{\frac{(1+i)^n-1-(1+i)^n}{(1+i)^n-1}\right\}$$

$$= P\left\{\frac{i(1+i)^n}{(1+i)^n - 1}\right\} - S\left\{\frac{i}{(1+i)^n-1}\right\}$$

(b) $[P - S(P/F,i\%,n)]\ (A/P,i\%,n)$

$$= P\left\{\frac{i(1+i)^n}{(1+i)^n-1}\right\} - S\left\{\frac{1}{(1+i)^n}\right\}\left\{\frac{i(1+i)^n}{(1+i)^n-1}\right\}$$

$$= P\left\{\frac{i(1+i)^n}{(1+i)^n-1}\right\} - S\left\{\frac{i}{(1+i)^n-1}\right\}$$

6-15 continued

(c) $Pi + (P-S)(A/F,i\%,n)$

$$= Pi + P\left\{\frac{i}{(1+i)^n-1}\right\} - S\left\{\frac{i}{(1+i)^n-1}\right\}$$

$$Pi\left\{1 + \frac{1}{(1+i)^n-1}\right\}$$

$$= Pi\left\{\frac{(1+i)^n-1+1}{(1+i)^n-1}\right\}$$

$$= P\left\{\frac{i(1+i)^n}{(1+i)^n-1}\right\} - S\left\{\frac{i}{(1+i)^n-1}\right\}$$

6-16

CR:	Type A	Type B
\$6,000(A/P,13%,12)-\$600(A/F,13%,12)		
=\$6,000(0.16899)-\$600(0.03899) =	\$991	
\$7,800(0.16899)-\$600(0.03899) =		\$1,295
Annual operation and repair costs	1,800	1,500
Extra annual income taxes		75
	\$2,791	\$2,870

6-17

CR:	Type P	Type Q
\$50,000(A/P,20%,12)=\$50,000(0.22526) =	\$11,263	
\$65,000(A/P,20%,20)-\$1,000(A/F,20%,20)		
=\$65,000(0.20536)-\$1,000(0.00536) =		\$13,343
Relining costs:		
\$6,000[(P/F,20%,4)+(P/F,20%,8)](A/P,20%,12)		
=\$6,000(0.4823+0.2326)(0.22526) =	966	
\$6,000(P/F,20%,10)(A/P,20%,20)		
=\$6,000(0.1615)(0.20536) =		199
Operation and maintenance		
\$2,400+\$150(A/G,20%,12)		
=\$2,400+\$150(3.484) =	2,923	
\$1,200+\$50(4.464) =		1,423
	\$15,152	\$14,965

6-18

	Type P	Type Q
CR:		
$50,000(A/P,20%,12)=$50,000(0.22526) =	$11,263	
$65,000(A/P,20%,12)-$1,000(A/F,20%,12)		
=$65,000(0.22526)-$1,000(0.02526) =		$14,617
Relining costs: (Same as 6-17)	966	199
Operation and maintenance:		
(Type P same as 6-17) =	2,923	
$1,200+$50(A/G,20%,12)		
=$1,200+$50(3.484) =		1,374
	$15,152	$16,190

6-19

	Plan I	Plan II
CR:		
$390,000(0.09)	$35,100	
$490,000(0.09)		$44,100
Annual maintenance:	12,000	4,000
Refurbishing costs:		
$50,000(A/F,9%,15)=$50,000(0.03406) =	1,703	
$100,000(A/F,9%,10)=$100,000(0.06582) =	6,582	
$50,000(A/F,9%,25)=$50,000(0.01181) =		591
$100,000(A/F,9%,20)=$100,000(0.01955) =		1,955
	$55,385	$50,646

6-20

The changes required to 6-19 include only the provisions for replacement of buildings and facilities. Calculation of the refurbishing costs accounts for $50,000 of the cost of replacing buildings and $100,000 of the cost of replacing facilities. This can be added to the previous costs as follows:

	Plan I	Plan II
From 6-19:	$55,385	$50,646
Buildings:		
$75,000(A/F,9%,75)=$75,000(0.00014) =	11	
$125,000(A/F,9%,75)=$125,000(0.00014) =		18
Facilities:		
$100,000(A/F,9%,40) = $100,000(0.00296) =	296	
$150,000(0.00296) =		444
	$55,692	$51,108

CHAPTER 7

Present Worth

7-1

	Plan C	Plan D
Initial investment	\$500,000	\$800,000
\$200,000(A/F,7%,20)/0.07		
=\$200,000(0.02439)/0.07 =	69,686	
\$30,000(P/F,7%,30)=\$30,000(0.1314) =		3,942
Annual disbursements:		
\$30,000/0.07-\$10,000(P/A,7%,20)		
=\$428,571-\$10,000(10.594) =	322,631	
\$10,000/0.07 =		142,857
	\$892,317	\$946,799

7-2

	Gas	Electric
Investment costs:	\$6,000	\$8,000
-\$1,000(P/F,10%,5)=-\$1,000(0.6209)		- 621
Annual disbursements:		
\$1,100(P/A,10%,5) = \$1,100(3.791) =	4,170	
\$800(3.791) =		3,033
	\$10,170	\$10,412

7-3

	Plan I	Plan II
Investment costs:	$50,000	$30,000
$25,000(P/F,9%,9)=$25,000(0.4604) =	11,510	
-$10,000(P/F,9%,18)=-$10,000(0.2120) =	- 2,120	
$30,000(P/F,9%,6)=$30,000(0.5963) =		17,889
$20,000(P/F,9%,12)=$20,000(0.3555) =		7,110
Annual disbursements:		
$11,000(P/A,9%,9)=$11,000(5.995) =	65,945	
$18,000[(P/A,9%,18)-(P/A,9%,9)]		
=$18,000(8.756-5.995) =	49,698	
$8,000(P/A,9%,6)=$8,000(4.486) =		35,888
$16,000[(P/A,9%,12)-(P/A,9%,6)]		
=$16,000(7.161-4.486) =		42,800
$25,000[(P/A,9%,18)-(P/A,9%,12)]		
=$25,000(8.756-7.161) =		39,875
	$175,033	$173,562

7-4

Semiannual interest = $10,000(0.05/2) = $250. Price now to yield a nominal 7%/annum, 3.5% effective semiannually:

PW = $250(P/A,3.5%,40)+$10,000(P/F,3.5%,40)
= $250(21.355)+$10,000(0.2526) = $7,865.

7-5

The investor can afford to pay the PW of the net receipts:

PW = $3,540(P/A,18%,15)+$25,000(P/F,18%,15)
= $3,540(5.092)+$25,000(0.0835) = $20,113

7-6

Two years have already elapsed. Thus we can assume that the first two $1,000 payments have already been made. The company can afford to pay the PW of the remaining 15 years' payments:

PW = $2,000(P/A,16%,15)+$3,000(P/A,16%,10)
- $4,000(P/A,16%,2) = $2,000(5.575)
+ $3,000(4.833) - $4,000(1.605) = $19,229.

7-7

The subsidy is the difference between the $2 million loan and the present worth of the $50,000 annual repayments discounted at 7% interest.

Subsidy = $2,000,000 - $50,000(P/A, 7%,40)
= $2,000,000 - $50,000(13.332) = $1,333,400.

7-8

The maximum price that the XYZ Company can affrod to pay for the property is the PW of the future royalties avoided and selling price of the land.

$$PW = \$2(20{,}000)(P/A,14\%,15) - \$1(20{,}000)(P/A,14\%,5) + \$30{,}000(P/F,14\%,15) = \$40{,}000(6.142) - \$20{,}000(3.433) + \$30{,}000(0.1401) = \underline{\$181{,}220}.$$

7-9

Since John Doe will have \$2,000/yr additional expenses and an i* of 11%, the PW of the future earnings to him is:

$$PW = \$38{,}000(P/A,11\%,15) - \$20{,}000(P/A,11\%,5) + \$30{,}000(P/F,11\%,15) = \$38{,}000(7.191) - \$20{,}000(3.696) + \$30{,}000(0.2090) = \underline{\$205{,}610}.$$

Doe and XYZ Company are pretty far apart in their views of the value of this property. Some irreducible factors Doe should consider are his declining health, his need for cash, alternative uses of the cash, the likelihood of renewal of the contract, the future of the XYZ Company, his relationship with his heirs, etc.

7-10

	Plan I	Plan II
Investment costs:	\$150,000	\$250,000
\$75,000(A/F,9%,30)/0.09		
=\$75,000(0.00734)/0.09 =	6,117	
\$120,000(A/F,9%,50)/0.09		
=\$120,000(0.00123)/0.09 =		1,640
Annual disbursements:		
\$7,000/0.09+\$3,000(P/A,9%,10)		
=\$7,000/0.09+\$3,000(6.418) =	97,032	
\$4,000/0.09 =		44,444
	\$253,149	\$296,084

7-11

	Type A	Type B
Investment costs:	\$135,000	\$100,000
\$15,000[(P/F,6%,20)+(P/F,6%,40)]		
=\$15,000(0.3118+0.0972) =	6,135	
\$20,000(0.3118+0.0972) =		8,180
\$80,000(P/F,6%,30)=\$80,000(0.1741)=		13,928
Annual disbursements:		
\$3,000(P/A,6%,60)+\$60(P/G,6%,60)		
=\$3,000(16.161)+\$60(239.043) =	62,826	
\$4,000(16.161)+\$80(239.043) =		83,767
	\$203,961	\$205,875

7-12

	Model H	Model K
Investment costs:		
$9,000+$6,500[(P/F,20%,3)+(P/F,20%,6)		
+(P/F,20%,9)]-$2,500(P/F,20%,12)		
=$9,000+$6,500(0.5787+0.3349+0.1938)		
-$2,500(0.1122) =	$15,918	
$11,000+$9,500[(P/F,20%,4)+(P/F,20%,8)]		
-$1,500(P/F,20%,12)=$11,000 +		
$9,500(0.4823+0.2326)-$1,500(0.1122) =		$17,623
Annual disbursements:		
[$2,200+$300(A/G,20%,3)](P/A,20%,12)		
=[$2,200+$300(0.879)](4.439) =	10,936	
[$1,700+$300(A/G,20%,4)](P/A,20%,12)		
=[$1,700+$300(1.274)](4.439) =		9,243
	$26,854	$26,866

7-13

	Unit A	Unit B
Investment costs:		
$40,000[1+(P/F,14%,8)+(P/F,14%,16)		
+(P/F,14%,24)+(P/F,14%,32)]		
=$40,000(1+0.3506+0.1229+0.0431+0.0151) =	$61,268	
$50,000+$45,000[(P/F,14%,10)+(P/F,14%,20)		
+(P/F,14%,30)]-$5,000(P/F,14%,40)		
=$50,000+$45,000(0.2697+0.0728		
+0.0196)-$5,000(0.0053) =		$66,268
Annual disbursements:		
$1,800(P/A,14%,40)=$1,800(7.105) =	12,789	
1,750(7.105) =		12,434
	$74,057	$78,702

7-14

	Unit X	Unit Y
Investment costs:		
$16,000-$2,000(P/F,20%,10)		
=$16,000-$2,000(0.1615) =	$15,677	
$24,000-$3,000(0.1615) =		$23,516
Annual disbursements:		
$9,000(P/A,20%,10)+$30(P/G,20%,10)		
=$9,000(4.192)+$30(12.887) =	38,115	
$7,650(4.192)+$15(12.887) =		32,262
	$53,792	$55,778

7-15

Investment costs:	Plan I	Plan II
	\$390,000	\$490,000
\$50,000(A/F,9%,15)/0.09 =\$50,000(0.03406)/0.09 =	18,922	
\$100,000(A/F,9%,10)/0.09 =\$100,000(0.06582)/0.09 =	73,133	
\$50,000(A/F,9%,25)/0.09 =\$50,000(0.01181)/0.09 =		6,561
\$100,000(A/F,9%,20)/0.09 =\$100,000(0.01955)/0.09 =		21,722
Annual disbursements:		
\$12,000/0.09 =	133,333	
\$4,000/0.09 =		44,444
	\$615,388	\$562,727

7-16

The scholarship will reach \$10,000 at the end of year 26. One way to set this up is as follows:

PW = \$5,000/0.08+\$200(P/G,8%,25)+\$5,000(P/F,8%,25)/0.08
= \$5,000/0.08+\$200(87.804)+\$5,000(0.1460)/0.08
= \$62,500+\$17,561+\$9,125 = \$89,186.

7-17

The scholarship will reach \$10,000 at the end of year 26 with the last award given at the end of year 30.

PW = \$5,000(P/A,8%,30)+\$200(P/G,8%,25)
= +\$5,000(P/A,8%,5)(P/F,8%,25)
= \$5,000(11.258)+\$200(87.804)+\$5,000(3.993)(0.1460)
= \$56,290+\$17,561+\$2,915 = \$76,766.

7-18

The scholarship will reach the \$10,000 level at the end of year 26. At the end of year 25, the amount remaining in the endowment fund will be:

F_{25} = \$80,000(F/P,8%,25) - \$5,000(F/A,8%,25)
- \$200(A/G,8%,25)(F/A,8%,25)
= \$80,000(6.8485)-\$5,000(73.106)
- \$200(8.225)(73.106) = \$547,880 - \$365,530
- \$120,259 = \$62,090.

The remaining period of the \$10,000 scholarship may be found by interpolating on n for the value of (A/P,8%,n):
(A/P,8%,n) = \$10,000/\$62,090 = 0.16105

7-18 continued

n is between 8 and 9. Thus, the last full $10,000 will be paid at the end of the 8th year. The 9th year scholarship to just exhaust the endowment will be:

[$62,090(F/P,8%,8)-$10,000(F/A,8%,8)](1.08)=
[$62,090(1.8509)-$10,000(10.637)](1.08) = $9,237.

Thus the full term of the scholarship will be 33 or 34 years depending upon how the final $9,237 scholarship is interpreted.

7-19

	Type X	Type Y
Investment costs:		
First cost:	$36,000	$44,000
Resurfacing: $18,000[(P/F,9%,10) +(P/F,9%,20)] = $18,000(0.4224+0.1784)	10,814	
$24,000(P/F,9%,15) = $24,000(0.2745)		6,588
Annual upkeep	3,400	1,800
	$50,214	$52,388

7-20

	System A	System B
Initial Investment:	$58,000	$72,000
Replacement: $16,000(P/F,11%,10) = $16,000(0.3522) =	5,635	
Annual costs: $12,000(P/A,11%,20) +$500(P/G,11%,20)=$12,000(7.963) +$500(49.842) =	120,477	
$10,000(P/A,11%,20)+$350(P/G,11%,20) =$10,000(7.963)+$350(49.842) =		97,075
	$184,112	$169,075

7-21

Present worth of the cash flows is: $8,300(P/A,18%,15) + $65,000(P/F,18%,15) = $8,300(5.092) + $65,000(0.0835) = $47,691. This is much less than the offering price so the investor should invest in the alternative offering 18% or more return.

7-22

Semi-annual interest payment is $100,000(0.07/2) = $3,500. The current sales price at 4.5% semiannually for 17 periods will be:

PW = $3,500(P/A,4.5%,17) + $100,000(P/F,4.5%,17)
= $3,500(11.707) + $100,000(0.4732) = $88,295.

7-23

Present - ($4,500+$3,000)[(P/F,15%,1)
+(P/F,15%,3)+(P/F,15%,5)
+(P/F,15%,7)+(P/F,15%,9)
+(P/F,15%,11)+(P/F,15%,13)]
=$7,500(0.8696+0.6575+0.4972
+0.3759+0.2843+0.2149+0.1625) = $22,964

Proposed- ($10,000+$3,000)[(P/F,15%,1)
+(P/F,15%,5)+(P/F,15%,9)
+(P/F,15%,13)]= $13,000(0.8696
+0.4972+0.2843+0.1625) = $23,576

7-24

With the shorter life, the relinings planned for year 13 will not be made in either case. This changes the present worth found in 7-23 to:

Present = $22,964 - $7,500(0.1625) = $21,745.
Proposed = $23,576 - $13,000(0.1625) = $21,464.

Where in 7-23, continuing with the present method was preferred by a PW of $612, with the shorter life, the proposed method is preferred by a PW of $281. Considering a lack of certainty with respect to the remaining life of the tank, the final decision should be based on factors which have not been reduced to money values.

7-25

	Structure C	Structure D
Investment costs:		
$26,000 + $21,000(P/F,9%,15)		
-$4,000(P/F,9%,30) = $26,000		
+$21,000(0.2745) - $4,000(0.0753) =	$31,463	
$50,000 - $5,000(P/F,9%,30)		
= $50,000 - $5,000(0.0753) =		$49,624
Annual costs: $4,800(P/A,9%,30)		
= $4,800(10.274) =	49,315	
$2,600(10.274) =		26,712
	$80,778	$76,336

7-26

Investment costs:	Plan X	Plan Y
$120,000[1+(P/F,10%,20)+(P/F,10%,40)]		
=$120,000(1+0.1486+0.0221) =	$140,484	
$185,000+$170,000(P/F,10%,30)		
-$15,000(P/F,10%,60)=$185,000		
+$170,000(0.0573) - $15,000(0.0033) =		$194,692
Annual disbursements & taxes		
$19,000(P/A,10%,60) = $19,000(9.967) =	189,373	
$14,000(P/A,10%,60) = $14,000(9.967) =		139,538
$2,667(9.967) =		26,582
	$329,857	$360,812

7-27

NPW = -$140,000 + $37,000(P/A,15%,10) + $500(P/G,15%,10)
+ [$41,200(P/A,15%,5) -$300(P/G,15%,5)](P/F,15%,10)
- $13,500(P/A,15%,15) - $200(P/G,15%,15)
= -$140,000 + $37,000(5.019) + $500(16.979)
+ [$41,200(3.352) - $300(5.775)](0.2472)
- $13,500(5.847)- $200(26.693) = +$3,630.

7-28

Semiannual interest payment on bond is
$10,000(0.07/2) = $350.

P = $350(P/A,i%/2,30) + $10,000(P/F,i%/2,30)
@6%: P = $350(19.600)+$10,000(0.4120) = $10,980
@8%: P = $350(17.292)+$10,000(0.3083) = $ 9,135
@9%: P = $350(16.289)+$10,000(0.2670) = $ 8,371
@10%: P = $350(15.372)+$10,000(0.2314) = $ 7,694

7-29

NPW = -$60,000+$8,000(P/A,i%,15)+$40,000(P/F,i%,15)
@ 6%; NPW = -$60,000+$8,000(9.712)+$40,000(0.4173) = $34,388.
@ 8%; NPW = -$60,000+$8,000(8.559)+$40,000(0.3152) = $21,080.
@10%; NPW = -$60,000+$8,000(7.606)+$40,000(0.2394) = $10,424.
@12%; NPW = -$60,000+$8,000(6.811)+$40,000(0.1827) = $ 1,796.
@15%; NPW = -$60,000+$8,000(5.847)+$40,000(0.1229) = -$8,308.

CHAPTER 8

Calculating An Unknown

Interest Rate

General Notes

It is immediately obvious that the determination of an unknown rate of return on an investment can be computed using either the present worth method or the equivalent uniform annual cost method. There is only one thing that must be carefully watched: viz. the present worth method requires that the same service for the same length of time be compared. Therefore, whenever the competing projects have different lives, it is usually easier to use the EUAC method to determine the unknown interest rate. The rate of return that makes the equivalent uniform cash flow of two projects (cash flows) equal is also the prospective rate of return on the extra investment of the project requiring the larger initial negative cash flow.

Linear interpolation between two interest rates will introduce some error in the answer. It is worth noting that the error will be in one direction if the PW method is used and in the opposite direction if the EUAC method is used. Thus, if the method used is opposite to the one used in this solution manual, the reader may find a difference in the answers in the second decimal place. In general, since most cash flows in typical problems are estimates of the future, the authors ordinarily advise against computing answers to more than one decimal place. In this solution manual, however, many of the solutions have been obtained by the use of a microcomputer that works all problems to six significant figures. Consequently, many answers are given to two or even three decimal places for the benefit of the instructor.

In previous chapters the authors have frequently omitted algebraic signs by identifying the cash flows as receipts or disbursements. In this chapter and in the solution manual the algebraic signs have been carefully indicated. The authors strongly advise students and instructors to use the proper signs in all solutions for several reasons. First, computers have come into common usage in most companies, universities and many homes. The computer requires that all equations include the proper signs in order to carry out the instructions. Secondly, many students use special hand-held calculators that have been programmed to solve engineering economy and financial problems. These calculators also require that proper signs be used. Finally, the convention of using the negative sign to indicate a disbursement and the positive sign to indicate a receipt is just good logic and their consistent use will assist students in understanding the process of economic analysis.

8-1

NPW = 0 = -\$100,000 +\$9,500(P/A,i%,5) +\$500(P/G,i%,5) +\$107,500(P/F,i%,5)
at i = 10%
NPW = -\$100,000 +\$9,500(3.791) +\$500(6.8618) +\$107,500(0.6209) = +\$6,192
at i = 12%,
NPW = -\$100,000 +\$9,500(3.605) +\$500(6.390) +\$107,500(0.5674) = -\$1,559
By interpolation, i = 10% + 2%(\$6,192/\$7,751) = 11.6%

8-2

NPW = 0 = -\$9,200 +\$300(P/A,i%,16) +\$10,000(P/F,i%,16)
At I = 3.5%, NPW = -\$9,200 +\$300(12.094) +\$10,000(0.5767) = +\$195.20
At i = 4%, NPW = -\$9,200 +\$300(11.652) +\$10,000(0.5339) = -\$365.40
By interpolation, i = 3.5% + 0.5(\$195.20/\$560.60) =3.674%
Nominal i = 2(3.674) = 7.348%
Effective i = $(1,03674)^2$ -1 = 7.48%

8-3

Yr.	Method A	Method B	(B - A)
0	-\$40,000	-\$55,000	-\$15,000
1	- 12,000	- 9,000	+ 3,000
2	- 14,000	- 10,000	+ 4,000
3	- 16,000	- 11,000	+ 5,000
4	- 16,000	- 11,000	+ 5,000
5	- 16,000	- 11,000	+ 5,000
5	0	+ 5,000	+ 5,000

NPW = 0 = -\$15,000 +\$3,000(P/F,i%,1) +\$4,000(P/F,i%,2) +\$5,000(P/A,i%,3)(P/F,i%,2) +\$5,000(P/F,i%,5)
At i= 18%, NPW = -\$15,000 +\$3,000(0.8475) +\$4,000(0.7182) +\$5,000(2.174)(0.7182) +\$5,000(0.44371) =+\$408
At i% = 20%, NPW = -\$15,000 +\$3,000(0.8333) +\$4,000(0.6944) +\$5,000(2.106)(0.6944) +\$5,000(0.4019) = -\$401
By interpolation, i =19.0%; Invest in B

8-4

NPW = 0 = +$9,300,000 -$540,000(P/A,i%,40) -$10,000,000(P/F,i%,n)

At i = 5.5%, NPW = +$9,300,000 -$540,000(16.046) -$10,000,000(0.1175) = +$539,840

At i = 6%, NPW = +$9,300,000 -$540,000(15.046) -$10,000,000(0.0972) = -$203,160

By interpolation, i = 5.863%

Nominal i = 11.726%

8-5

NPW = 0 = -$40,000 +($8,400-$3,000)(P/A,i%,1z) -($500-$250)(P/G,i%,15) +$4,000(P/F,i%,15)

At i = 5%, NPW = -$40,000 +$5,400(10.380) -$250(63.2880) +$4,000(0.4810) = +$2,154

At i = 6%, NPW = -$40,000 +5,400(9.712) -$250(57.5546) +$4,000(0.4173) = -$274.65

By interpolation, i = 5.887%

8-6

Since Plan I has a 30-year life and Plan II has a 50-year life, the best solution is to equate the EUAC of Plan I to the EUAC of Plan II at unknown interest rate, i. If i is equal to or greater than i*, select Plan II.

-$150,000(A/P,i%,30) +$75,000(A/F,i%,30) -$7,000 -$3,000(P/A,i%,10)(A/P,i%,30) = -$250,000(A/P,i%,50) +$130,000(A/F,i%,50) -$4,000

At i = 5%, EUAC Plan I = -$150,000(0.06505) +$75,000(0.01505) -$7,000 -$3,000(7.722)(0.06505) = -$17,136

EUAC Plan II = -$250,000(0.05478) +$130,000(0.00478) -$4,000 = -$17,074

At i = 5.5%, EUAC Plan I = -$17,842

EUAC Plan II = -$18,237

By interpolation, i = 5.07%

8-7

Equate PW of 60 years of service at unknown i:

Type A : NPW = -$120,000 - $15,000(1+ (P/F,i%,20) +(P/F,i%,40)) -$3,000(P/A,i%,60) -$60(P/G,i%,60)

Type B : NPW = -$80,000(1 + (P/F,i%,30)) -$20,000(1 +(P/F,i%,20) +(P/F,i%,40)) -$4,000(P/A,i%,60) -$80(P/G,i%,60)

	i=6%	i=7%
NPW Type A	-$203,961	-$193,141
NPW Type B	- 205,875	- 188,033
(B- A)	+ 1,914	- 5,108

i = 6% +1%($1,914/$7,022) = 6.27%

8-8

Yr.	Machine K	Machine J	(J-K)
0	-$30,000	-$50,000	-$20,000
1	-8,000	-6,000-750	+1,250
2	-8,500	-6,300-850	+1,350
'			
'	G=-$500	G=-$400	G= +$100
12			
12	0	+$14,000	+$14,000

NPW = 0 = -$20,000 +$1,250(P/A,i%,12) +$100(P/G,i%,12)
+$14,000(P/F,i%,12)
Try 6% and 7%: i = 6.92% i*; Select Machine K

8-9

NPW =0= -$18,000 +$1,000(P/A,i%,2)+$5,000(P/A,i%,8)(P/F,i%,2)
+$2,000(P/A,i%,5)(P/F,i%,10)
At i = 18%, NPW = -$18,000 +$1,000(1.566)
+$5,000(4.303)(0.7182) +$2,000(3.127)(0.1911)
+$213.213
At i = 20%, NPW = -$18,000 +$1,000(1.528) +$5,000(3.837)
(0.6944) +$2,000(2.991)(0.1615) = -$2,183.84

By interpolation, i= 18.18%, greater than i*. Buy patent.

8-10

Yr.	Gas	Electric	(E-G)
0	-$6,000	-$8,000	-$2,000
1 - 5	- 1,100	- 800	+ 300
5	0	+ 1,000	+ 1,000

NPW = 0 = -$2,000 +$300(P/A,i%,5) +$1,000(P/F,i%,5)
At i = 6%, NPW = +$10.97
At i = 7%, NPW = -$56.95
By interpolation, i = 6.16%

8-11

Yr.	Project Y	Project Z	(Z-Y)
0	-$100,000	-$140,000	-$40,000
1- 20	+ 15,000	+ 17,750	+ 2,750

Project Y: NPW = 0 = -$100,000 +$15,000(P/A,i%,20)
(P/A,i%,20) = 6.6667; i =13.9%
Project Z NPW = 0 = -$140,000 +$17,750(P/A,i%,20)
(P/A,i%,20) = 7.88732; i = 11.2%
i on extra investment in Z over Y:
NPW = 0 = -$40,000 +$2,750(P/A,i%,20)
(P/A,i%,20) = 14.5454; i = 3.3%; Do not select Z.

8-12

NPW = 0 = -\$100,000 -\$1,200(P/A,i%,5) +(\$250,000 - \$16,000)(P/F,i%,5)
Try i = 16% and 18%
By interpolation, i = 17.7%

8-13

NPW = 0 = -\$120,000 +\$40,000(P/F,i%,10) +(\$30,000-\$15,000) (P/A,i%,10)
Try i = 7% and 8%; By interpolation, i = 7.78%, less than i*; Reject Proposal.

8-14

NPW = 0 = -\$120,000 +\$40,000(P/F,i%,10) +\$12,000(P/A,i%,10) +\$1,000(P/G,i%,10)
Try 8% and 9%;
i = 8.74%, less than i* ; Reject proposal.

8-15

NPW = 0 = -\$850 +\$37.50(P/A,i%,12) +\$1,000(P/F,i%,12)
i = 5.49%
Nominal i = 2(5.49%) = 10.98%
Effective i = $(1.0549)^2 - 1 = 11.28\%$

8-16

If the Jones build on the lot:
NPW = 0 = -\$25,000 -\$300(P/A,i%,4) +\$38,000(P/F,i%,4)
i = 10%; This is an attractive investment for the Jones.
If they sell the lot:
NPW = 0 = -\$25,000 -\$300(P/A,i%,4) +\$35,400(P/F,i%,4)
i = 8.03%; This is borderline, and irreducibles will probably be the basis for the decision since the tax exempt bonds pay about 8.5%

8-17

NPW = 0 = -\$185,000 +\$18,000(P/A,i%,15) +\$20,000(P/A,i%,10) (P/F,i%,5) +\$25,000(P/F,i%,15)
At i = 12%, NPW = +\$6,282
At i = 13%, NPW = -\$5,782; i = 12.52%.

In Problem 7-9 Doe's i* was stated as 11%. If that is all he can expect to earn on the \$185,000 he would receive from the sale of the land, then he is better off to keep it. Some irreducible factors he should consider are his declining health, his need for cash, alternative uses of the cash, the likelihood of the company renewing the contract and the future of the XYZ Company.

8-18

NPW = 0 = -$200,000 +($32,000 -$12,400)(P/A,i%,20) +$180,000(P/F,i%,20)

Try i = 9% and i = 10%; i = 9.62%

8-19

NPW in 1972 = 0 = -$5,300 +$318(P/A,i%,3) +$480(P/A,i%,5) (P/F,i%,3) +$9,600(P/F,i%,8)

At i = 13%, NPW = +$232.38

At i = 14%, NPW = -$83.55; i =13.74%

8-20

Dealer's Proposal:

NPW = 0 = +$5,925 -$216.91(P/A,i%,36) -$75((P/F,i%,12) + (P/F,i%,24))

Try i = 1.5%; by interpolation, (P/A,1.5%,36)= 27.644

NPW = +$5,925 -$216.91(27.644) -$75(0.8364 +0.6995) = -$186.45

Try i = 2%; by interpolation, (P/A,2%,36) = 25.4702

NPW = +$5,925 -$216.91(25.4702) -$75(0.7885 + 0.6217)= +$294.49

By interpolation, i = 1.694%

Effective i = $(1.01694)^{12}$-1 = 22.3%

Bank's proposal:

NPW = 0 = +$6,000 -$207.99(P/A,i%,36)

(P/A,i%,36) = ($6,000/$207.99) = 28.8475

By interpolation, i = 1.2502%

Effective i = $(1.012502)^{12}$ -1 = 16.08%

8-21

Yr.	Toltec	Mandan	(T-M)
0	-$5,000	-$3,800	-$1,200
1-25	- 3,242	- 3,340	+ 98

NPW = 0 = -$1,200 +$98(P/A,i%,25)

(P/A,i%,25) = ($1,200/$98) = 12.2449

By interpolation, i = 6.476%

8-22

NPW = 0 = +$19,200 -$400 -$2,250(P/A,i%,20) -$20,000(P/F,i%,20)

Try i = 12%; NPW = +$79.574

Try i = 13%; NPW = -$125.867

By interpolation, i = 12.39% (Note, if solved by the microcomputer, the true i = 12.056%. This shows the magnitude of the error due to linear interpolation.)

8-23

Yr.	Project X	Project Y	(Y-X)
0	-$250,000	-$325,000	-$75,000
1-25	+ 32,000	+ 36,500	+ 4,500
25	+ 50,000	+ 50,000	0

Project X: NPW = -$250 +$32(P/A,i%,25) +$50(P/F,i%,25)
i = 12.2%, greater than i*. Project X is acceptable.
Rate of return on extra investment in Project Y = i;
NPW = -$75 +$4.5(P/A,i%,25)
(P/A,i%,25) = ($75/$4.5) = 16.6667
By interpolation, i = 3.4%, less than i*. Reject Project Y.

8-24

NPW = 0 = -$10,000 +$500(P/A,i%,5) +$400(P/A,i%,5)(P/F,i%,5) +$12,000(P/F,i%,10)
At i = 6%, NPW = +$65.85
At i = 7%, NPW = -$681.08; i = 6.09%

8-25

Yr.	Unit X	Unit Y	(Y-X)
0	-$16,000	-$24,000	-$8,000
1	- 9,000	- 7,650	+ 1,350
2	- 9,030	- 7,665	+ 1,365
'			
'	G = -30	G = -15	G = +15
10			
10	+	+ $3,000	+$1,000

NPW = -$6,000 +$1,350(P/A,i%,10) +$15(P/G,i%,10) +$1,000(P/F,i%,10)
At i = 12%, NPW = +$253.59
At i = 13%, NPW = -$ 93.79; i = 12.73%

8-26

(a) 1-year subscription vs news stand price:
NPW = 0 = -$21 +$2(P/A,i%,12); i = 2.1166
Effective i = 28.6%
(b) 2-year subscription vs news stand price:
NPW = 0 = -$38 +$2(P/A,i%,24); i = 1.9600%
Effective i = 26.2%
(c) 3-year subscription vs news stand price:
NPW = -$52 +$2(P/A,i%,36); i = 1.8769
Effective i = 18.8%
(d) 2-year subscription vs 1-year subscription:
NPW = 0 = -$17 +$21(P/F,i%,1); i = 23.5%
(e) 3-year subscription vs 1-year subscription:
NPW = 0 = -$31 +$21(P/A,i%,2); i = 22.8%

8-27

Yr.	K-400	E-450	(E-K)
0	-$4,000	-$8,000	-$4,000
1 - 4	- 1,850	- 1,400	+ 450
4	- 3,000	0	+ 3,000
5 - 8	- 1,850	- 1,400	+ 450
8	+ 1,000	+ 2,000	+ 1,000

NPW = 0 = -$4,000 +$450(P/A,i%,8) +$3,000(P/F,i%,4)
+$1,000(P/F,i%,8)
At i = 15%, NPW = +$65.50
At i = 16%, NPW = -$83.30
i = 15.4%, greater than i*;Select E-450.

8-28

Month	Purchase	Extra Income Tax	Lease	(P-L)
0	-$7,000		-$200	-$6,800
1-35	0		- 200	+ 200
12		-$324		- 324
24		- 324		- 324
36		- 324		- 324
36	+ 1,000			+ 1,000

(a) Before Income tax:
NPW = 0 = -$6,800 +$200(P/A,i%,35) +$1,000(P/F,i%,36)
i = slightly less than 1% by computer. Effective rate = 11.8%
(b) After Income Tax:
NPW = 0 = -$6,800 +$200(P/A,i%,35) +$1,000(P/F,i%,36)
-$324((P/F,i%,12) + (P/F,i%,24) +(P/F,i%,36))
By computer, i = 0.2317%; Effective rate = 2.81%
(c) Leasing Company:
NPW = 0 = -$6,200 +$200(P/A,i%,35) +$1,400(P/F,i%,36)
i = 1.62%; Effective i = 21.2%

8-29

Yr	Buy	Make (after tax)	(M-B) Before Tax	(M-B) After Tax
0	0	-$25,000	-$25,000	-$25,000
1	-$10,000	- 5,000	+ 6,000	+ 5,000
2	- 11,000	- 5,750	+ 6,500	+ 5,250
'	G = -1,000	G = -750	G = +500	G = +250
5				
5	0	+ 5,000	+ 5,000	+ 5,000

(a) Before Tax
NPW = 0 = -$25,000 +$6,000(P/A,i%,5) +$500(P/G,i%,5)
+$5,000(P/F,i%,5)
Try i = 15% and i = 16% : i = 15.7%

(b) After Tax:

NPW = 0 = -\$25,000 +\$5,000(P/A,i%,5) +\$250(P/G,i%,5) +\$5,000(P/F,i%,5)

Try i = 8% and i = 9%: i = 8.28%

Since i* = 9% after tax, this is a borderline proposal. Decision will probably be based on irreducibles, such as control of supplies and schedules, less risk of interruption of supply, purchase price may increase more rapidly than estimated. Conversely, the need for the part may not last for 5 years, or demand may decrease.

CHAPTER 9

Benefit-Cost Ratio

General Notes

The problems of Chapter 9 are designed to reinforce the analysis of the Benefit-Cost Ratio as a method for decision making in public projects. They help to bring out the fundamental fact that a great many of the benefit-cost analysis that have been used by government agencies have been incorrect because the incremental benefit - incremental cost approach has not been used. They also help to emphasize the fact that the interpretation of the benefit-cost ratio is frequently in error. The choice of the project that gives the highest ratio may or may not be the best, and it depends on a number of other factors in the situation and the alternatives available.

These ideas will become much more important when the student studies Chapter 19. Some Aspects of Economy Studies for Government Activities, and the teacher may be able to stimulate the students to think in terms of allocation of resources and the effects of using erroneous benefit-cost methods.

9-1

The following calculations use the equivalent uniform annual cost method and the data from Table 9-2. The same answers will be obtained using the present worth method and the information from Table 9-1. (000) omitted in all the money amounts.

For J over H, $\frac{B}{C} = \frac{(\$266.5 + \$35) - (\$199.9 + \$21)}{\$58.8 - \$10.4} = \frac{\$80.6}{\$48.4} = \underline{1.67}$

For K over J, $\frac{B}{C} = \frac{(\$199.9 + \$21) - (\$173.2 + \$17)}{\$109.3 - \$58.8} = \frac{\$30.7}{\$50.5} = \underline{0.61}$

For K over H, $\frac{B}{C} = \frac{(\$266.5 + \$35) - (\$173.2 + \$17)}{\$109.3 - \$10.4} = \frac{\$111.3}{\$98.9} = \underline{1.13}$

9-2

(000) omitted in all money amounts.
For D& R over CI, B/C = \$50/(\$157 + \$38) = \$50/\$190 = 0.26
For D&R over NFC: B = \$480 - \$55 = \$425;
C = \$5,300(A/P,6%,50)+\$40+\$38 = \$5,300(0.06344) + \$78
= \$414; B/C = \$425/\$414 = 1.03.

9-3

(Solutions to 9-3 and 9-4 illustrate slightly different methods of handling gradient series.) (000) omitted from all money amounts.

For Location H

Investment	\$110
PW of maintenance costs = \$35(P/A,7%,20)=\$35(10.594) =	371
PW of road user costs = \$300(P/A,7%,20)-\$90(P/A,7%,10) +\$10(P/G,7%,10) = \$300(10.594)-\$90(7.024) +\$10(27.716) =	2,823
	\$3,304

For Location J; Investment minus residual value =

\$700-\$300(P/F,7%,20)=\$700-\$300(0.2584) =	\$ 622
PW of maintenance costs=\$21(P/A,7%,20)=\$21(10.594) =	223
PW of road user costs=\$225(P/A,7%,20)-\$67.5(P/A,7%,10) +\$7.5(P/G,7%,10)=\$225(10.594)-\$67.5(7.024) +\$7.5(27.716) =	2,117
	\$2,962

For Location K; Investment minus residual value=

\$1,300 - \$550(0.2584) =	\$1,158
PW of maintenance costs = \$17(10.594)=	180
PW of road user costs = \$195(10.594)-\$58.5(7.024) +\$6.5(27.716) =	1,835
	\$3,173

9-4

(000) omitted from money amounts.

For Location H

CR = \$110(A/P,7%,20) = \$110(0.09439) =	\$10.4
Annual maintenance costs =	35.0
EUA road user costs = \$300 -[\$90(P/A,7%,9) -\$10(P/G,7%,9)](A/P,7%,20)=\$300-[\$90(6.515) -\$10(23.140)](0.09439) =	266.5
	\$311.9

9-4 continued

For Location J

CR = \$700(A/P,7%,20)-\$300(A/F,7%,20) =	
\$700(0.09439)-\$300(0.02439) =	\$58.8
Annual maintenance costs =	21.0
EUA road user costs = \$225-[\$67.5(6.515) -\$7.5(23.140)](0.09439) =	199.9
	\$279.1

For Location K

CR = \$1,300(0.09439)-\$550(0.02439)	\$109.3
Annual maintenance costs =	17.0
EUA road user costs = \$195-[\$58.5(6.515) -\$6.5(23.140)](0.09439) =	173.2
	\$299.5

9-5

For CI over NFC, an extra investment of \$2,900,000 causes annual net favorable consequences of (\$430,000-\$105,000)-\$35,000 = \$340,000.
(A/P,i%,50) = \$340/\$2,900 = 0.11724;
by interpolation, i = 11.7%

For D&R over CI, an extra investment of \$2,400,000 causes annual net favorable consequences of (105,000 - \$55,000) -(\$28,000 + \$10,000) - (\$40,000 - \$35,000) = \$7,000.

It is clear that the prospective rate of return is negative for a project with a 50-year life and zero terminal salvage value. It would take 343 years for an initial investment of \$2,400,000 to be recovered by \$7,000 a year, even if i = 0%. Even if the project were assumed to have perpetual life, the prospective i would be only 0.3%.

9-6

Annual highway costs for M: (000) omitted.

CR = \$3,000(A/P,8%,20) - \$1,800(A/F,8%,20)	
= \$3,000(0.10185)-\$1,800(0.02185) =	\$266.22
Operataion & maintenance	120.00
	\$386.22

Annual highway costs for N:

CR = \$5,000(0.10185) - \$3,000(0.02185) =	\$443.70
Operation & maintenance =	90.00
	\$533.70

B/C = (\$880-\$660)/(\$533.70-\$386.22) = 1.49

9-7

B/C ratios for the four schemes of classification are: (000) omitted.

(a) (\$1,100 - \$300)/(\$300 + \$700) = 0.80
(b) \$1,100/(\$300 + \$700 + \$300) = 0.85
(c) (\$1,100 - \$300 - \$700)/(\$300) = 0.33
(d) (\$1,100 - \$700)/(\$300 + \$300) = 0.67

The second classification, which classifies the \$300,000 as a cost, gives the most favorable ratio. The third scheme, which classifies O&M expenses as a "disbenefit", gives the lowest ratio

9-8

CI vs NFC. (000) omitted from all money amounts.

C = \$400(A/P,9%,30)+\$30 = \$400(0.09734) + \$30 = \$68.9
B = (\$300-\$180) + (\$20-\$12)(A/G,9%,30)
= \$120 + \$8(8.666) = \$189.3; B/C = 2.75.

D&R vs CI.

C = \$1,600(0.09734) + \$45 = \$200.7
B = (\$180-\$30) + (\$12-\$5)(8.666)-\$20 + \$8 = \$198.7
B/C = 0.99.

CI + D&R vs NFC.

C = \$2,000(0.09734) + \$75 = \$269.7
B = (\$300-\$30) + (\$20-\$5)(8.666)-\$20 + \$8 = \$388.0
B/C = 1.44.

Additional benefits do not exceed additional costs for D&R in addition to CI although it is a very close call.

9-9

NFC. (000) omitted from all money amounts.

PW = \$300(P/A,9%,30) + \$20(P/G,9%,30) = \$300(10.274) + \$20(89.028) = \$4,863.

CI. PW = \$400 + \$30(10.274)+\$180(10.274)+\$12(89.028)
= \$3,626.

CI + D&R PW = \$2,000+(\$75+\$30)+\$20-\$8)(10.274)+\$5(89.028)
= \$3,647.

9-10

CI vs NFC. (000) omitted from all money amounts.

C = $400(A/P,5%,30) + $30 = $400(0.06505) + $30 = $56.0;
B = ($300-$180) + ($20-$12)(A/G,5%,30) = $120+$8(10.969)
= $207.8; B/C = 3.71.

D&R vs CI.

C = $1,600(0.06505) + $45 = 149.1
B = ($180-$30) + ($12-$5)(10.969) - $20 + $8 = $214.8
B/C = 1.44.

CI + D&R vs NFC.

C = $2,000(0.06505)+ $75 = $205.1
B = ($300-$30) + ($20-$5)(10.969) - $20 + $8 = $422.5
B/C = 2.06.

9-11

Considered by itself as an addition to the D&R project;
C = ($100-$15)(A/P,9%,30) + $5 = $85(0.09734)
+$5 = $13.3; B = $18; B/C = 1.36.
However, the D&R project did not have a B/C ratio greater than 1. Therefore, D& R plus this modification should be considered together. D&R modified vs CI:
C = $1,685(0.09734) + $50 = $214.0;
B = ($180-$30)+($12-$5)(8.666)-$20 + $8 + $18 = $216.7;
B/C = 1.01. This modification changes the D&R project from being barely unacceptable to barely acceptable. The overall B/C ratio for CI and modified D&R will be:
C = $2,085(0.09734) + $80 = $283.0; B = ($300-$30)
+ ($20-$5)(8.666) - $20 + $8 + $18 = $406.0;
B/C = 1.43.

9-12

Annual highway costs are the same as in Problem 9-6, i.e., (000) omitted, $386.22 for M and $533.70 for N.
B = ($880-$660) + ($30-$25)(A/G,8%,20) = $220
+ $5(7.037) = $255.2
B/C = $255.2/($533.70 - $386.22) = 1.73
Route N is preferred to M.

9-13

Since bids A & B essentially afford the same protection, a choice between them may be made based on minimum Present Worth of costs.

PW(A) = \$125,000 + \$25,000[(P/F,9%,5) + (P/F,9%,10)] + \$15,000(P/A,9%,15) = \$125,000 + \$25,000(0.6499+0.4224) + \$15,000(8.061) = \$272,723.

PW(B) = \$160,000 + \$12,000(8.061) = \$256,732.

B is preferred to A. In comparing C to B, it must be remembered that C offers 15% improvement in efficiency.

PW(C) = \$190,000 + \$9,200(8.061) = \$264,161.

The 15% improvement in efficiency can be obtained at an equivalent of an additional PW of \$7,429.

Factors the community should consider in a final decision include possible future changes in requirements and the value of a cleaner river. It may want to go into further negotiation with bidders B and C.

9-14

PW(A) = \$125,000 + \$25,000[(P/F,4%,5) + (P/F,4%,10)] + \$15,000(P/A,4%,15) = \$125,000 + \$25,000(0.8219+0.6756) + \$15,000(11.118) = \$329,208.

PW(B) = \$160,000 + \$12,000(11.118) = \$293,416.

PW(C) = \$190,000 + \$9,200(11.118) = \$292,286.

In this case, Bid C is clearly preferred. The low interest rate encourages greater initial investment and, in this case, favors investment in a more efficient system.

9-15

Government annual highway costs are (000 omitted):

Route M \$386.22; Route N \$533.70 (See Problem 9-6).

Annual road user costs (000 omitted):

Route M: \$800 + \$15(A/G,8%,20) = \$880 + \$15(7.037) = \$985.6

Route N: \$660 + \$11.5(7.037) = \$740.9

B/C = (\$985.6 - \$740.9)/(\$533.70 - \$386.22) = 1.66.

CHAPTER 10

Choice of A Minimum

Attractive Rate of Return

General Notes

Chapter 10 is an explanation of a simple concept,, but one that has been badly confused in the public press, in theoretical economic papers, and by "model builders" who find it convenient to assume that any amount of money is always available for investment if one is willing to borrow.

The United States Government which has one of the highest credit ratings of any institution in the world has recently found it necessary to pay unheard of interest rates (such as 15.8% on six month treasury bills) in order to raise the funds to meet its obligations. It is competing in the money market for investment funds and as a result of these high interest rates, many industries such as housing, plant expansion, automobiles and durable goods are in a mild recession. This is a simple illustration of the fact that every organization has some kind of a limit to the availability of resources and that every organization should try to allocate its resources in a way so as to maximize the benefit derived from the investments. Consequently, the concept of opportunity cost as a "floor" or minimum rate of return to use in project evaluation and the ideas of capital rationing based on some minimum acceptable rate of return as a means of selecting competing proposals are valid. Appendic C clearly shows that the "reinvestment" theorist introduces a "red herring" into the logic of decision making about proposed investments.

It is important for the teacher to emphasize the fact that most of the problems described in this book deal with the choice among alternative ways of doing something or providing a service. The great bulk of the decisions are made at the "design" level, where a decision is made as to which of several possible proposals for doing the same thing will be presented to top management to compete with other proposals, similarly chosen, for doing many other things. The designers, draftsmen, production engineers, city engineers, etc., do not and cannot know what other opportunities are available to top management. They must be given specific instructions by top management in order to have some basic criteria by which they can judge the relative attractiveness of mutually exclusive competing proposals. The specification of a minimum attractive rate of return, either before or after income taxes is the best way to convey one management criterion to the project analysts.

In addition to the emphasis throughout this book on the analysis of the monetary differences among alternative proposals, there is also the emphasis, albeit not as strong, on the importance of the tabulation and consideration of irreducibles in arriving at a decision. This chapter attempts to illustrate through examples and problems the fact that the selection of the minimum attractive rate of return can be influenced by the individual's appraisal of the risk involved, personal opinions regarding accuracy of estimates, and kowledge of the technology involved. Thus, several persons reviewing the same data may arrive at several different rates to be used. This is really the result of applying irreducibles to the mathematical analysis.

<u>10-1</u>

Project	Life Years	Rate of Return %	Investment	Cumulative Investment
M	10	23.01	$ 25,000	$25,000
0	15	13.77	25,000	50,000
L	10	12.42	25,000	75,000
N	5	10.70	50,000	125,000
P	10	10.48	25,000	150,000
Q	12	8.91	50,000	200,000
K	8	7.33	50,000	250,000

(a) 10.7%. Invest in projects M, 0, L and N.

(b) This will change the order of ranking but not the projects selected. Projects L and 0 will be reversed.

<u>10-2</u> There is no finite solution to this problem. The economic conditions existing at the time the problem is assigned will affect the trend of the discussion. Some specific factors which might be included in the discussion are :

What is the "prime rate" banks are charging their best customers?

Are interest rates high or low compared to rates over the last several years ?

What effects of the current trends in interest rates on prices of bonds are discernable?

Are tax-exempt municipal bonds earning more or less than their stated interest rates ? Why ?

Are current interest rates reflected in stock market prices? How and why ?

It is interesting to have the students review the earning rates of large corporations over a period of years by having them examine several successive annual reports or by looking up the earnings rates in several successive FORTUNE magazine's annual reports on the "500" largest firms in the USA. Most students will be quite surprized at how low the actual earnings are, but they should compare them with the going interest rates of the same periods of time in order to obtain an insight into the performance of the corporations.

<u>10-3</u> It is suggested that some source document, such as the U. S. Department of Commerce's Business Statistics, be used for this problem. In addition to the rates of return indicated at the time of issue, the current earning or yield computed from current market price and the prospective receipts if held to maturity can be determined for a number of high grade and medium grade stocks and bonds.

<u>10-4</u> Mr. and Mrs. Smith must recognize that their time is valuable, worth $33,000 to someone else, in fact. Their net income from both their time and their investment is only $35,000. Therefore, the difference of $2,000 is all they can attribute to the effect of their investment.

(a) If their choice is between owning the franchise and working for someone else, what return can they obtain on their $18,000 in some equally secure investment ? Check the current rates on savings accounts, T-Bills, and Money Market Certificates. These rates will represent the opportunitiesavailable to the Smiths.

How good is the investment in the franchise ? What will its salvage value be in the future ? Furniture and equipment will need to be replaced about every 10 years. The restaurant will have to be redecorated periodically. The real value of the franchise will depend to a great extent on the quality of the food and service and the neighborhood. It is not unreasonable to assume a zero salvage value after 10 years. Thus,

(A/P,i%,10) = $2,000/$18,000 = 0.11111

i = 1.96%; This is not a very attractive investment because the Smiths can make any number of very secure investments at a much higher rate.

Suppose a 20-year life is assumed:

(A/P,i%,20) = $2,000/$18,000 = 0.11111

i = 9.1%; This looks considerably more attractive.

(b) For a 10-year life, (A/P,i%,10) = $2,000/$15,000 = 0.13333; i= 5.6%

For a 20-year life, (A/P,i%,20) = 0.13333; i = 11.93%

This looks a little better, but probably is not attractive in comparison with other investment opportunities.

Another approach might be to assume that some amount, perhaps $10,000, was for the franchise alone and the remainder, $5,000 or $8,000, was for equipment that would have a zero salvage value in X years. Assuming the franchise will always be worth $10,000 will make a big difference in the attractiveness of the two parts of this problem.

10-5

Proposal	Life Years	Rate of Return %	Investment	Cumulative Investment
B	7	20.0	$70,000	$70,000
G	20	19.0	80,000	150,000
A	10	18.2	50,000	200,000
F	14	18.0	60,000	260,000
D	6	16.9	40,000	300,000
C	4	16.0	20,000	320,000
E	8	16.0	50,000	370,000

(b) i* = 18.2%

(c) Proposals A and F will be preferred to Proposal G. Thus, B, A, and F could be selected, using a total of only $180,000, however Proposals C and D are equally attractive as F. Thus, the selection might be B, A, F, and C, giving an i* = 16% with the maximum life being 14 years for F.

10-6 The student should find out what the going rates are on T-Bills, Savings and Loan time deposits, bank time deposits, bank savings accounts (pass book accounts) and money market funds. He should evaluate the risk involved in each and determine whether or not he is willing to take that risk.(For example the money market funds are not insured by a Federal agency.) Then he should plan his investments so that he can maximize his return. Without going into money market certificates, he probably can keep out about half of his annual needs and invest that in a pass book account so that it is readily available. The remainder of his first year needs ($3,750) should go into a 6-month money market or S & L, or time deposit, which ever pays a higher rate.The balance, $12,500, can be invested in 1-year and 2-year time deposits or in T-Bills (26 weeks), which ever pays a higher rate. The $10,000 minimum investment in T-Bills could pose a difficulty after 1.5 years.

Assuming a 6-month time deposit rate of 7% and T-Bill rate of 12% it can be shown that the student can have his necessary $7,500 each year and still have a bit left over when he makes his last withdrawal 2.5 years hence.

10-7 The letter should note the current interest rates being charged for home mortgages, automobiles, and durable goods bought on time payment plans. It should recognize that interest rates vary over time and take into account the interest rate changes over the last few years. Especially, the existence of variable rate home mortgages and the history of the rates over a few years should be noted.

The letter should point out that government use of tax income for investments that return less than these going rates means that the tax payer is subsidizing projects for the government when he could use the money more effecively to reduce his own debts and interest payments--completely risk-free investments.

The teacher should not be surprized to find a wide variation in the attitudes of the students toward this problem. Some will consider low interest rates for government projects as being conservative. Some will try to justify different i*s for government projects based on their purposes. For example, they may propose low rates for social programs and high rates for defense projects. Some may even consider redistribution of wealth a justification for low i*s.

CHAPTER 11

Some Relationships Between Depreciation Accounting and Engineering Economy

General Notes

Depreciation accounting is a process of allocating to each year's operations a portion of the original investment in depreciable assets, and the charge represents a reduction in the taxable income on which income taxes would otherwise be paid. Engineering economy normally deals with the future and the consequences of an investment. Thus, the engineering economist must be familiar with depreciation accounting practices in order to predict the effects of the investment and the resulting depreciation on the income taxes and other expenses. It is important for the teacher to convey the concept to the students so they can always distinguish the significant differences between writing a financial history of an organization and the prediction of the effects of a decision on the financial health of the organization.

From 1954 to 1981 a choice has usually been open to the accountant as to how he or she will account for the decrease in value of an asset through the choice of a depreciation method to be used. In 1981 the Economic Recovery Tax Act, passed by the U. S. Congress, made significant changes in the laws regarding depreciation accounting for depreciable assets acquired after January 1, 1981. Thus, assets acquired before that date will continue to be treated under the older laws, while the new assets may be treated under the new law. Students will need to understand both sets of laws. Since the Recovery Tax Act was passed after the manuscript for the textbook was set in type, a new Appendix, Appendix F, was added to describe the most important aspects of the new law. It is suggested that Appendix F be assigned at the same time that Chapter 11 is assigned and that a little more time be allocated to the subject of depreciation accounting.

11-1 P = \$18,000, L = \$3,000, n = 5, i = 8%

(a) S.L. : D_1 = (\$18,000 - \$3,000)/5 = \$3,000

(b) D.D.B.: D_1 = (\$18,000)(2.0/5) = \$7,200

(c) SOYD: (5 x 6)/2 = 15 ; D_1 = 5(\$15,000/15) = \$5,000

(d) S.F. : D_1 = \$15,000(A/F,8%,5) =\$15,000(0.17046)
= \$2,557

11-2 From Problem 11-1:
(a) S.L.: BV_3 = \$18,000 - 3(\$3,000) = \$9,000

(b) D.D.B.: BV_1 = \$18,000(1 - 0.4) = \$18,000(0.6)
= \$3,888

(c) SOYD : BV_3 = \$18,000 - \$1,000(5 + 4 + 3)
= \$6,000

(d) S.F. : BV_3 = \$18,000 - \$2,557(F/A,8%,3) = \$9,700

11-3 P = \$24,750; L = 0; n = 10
(a) SOYD: SOYD = (10 x 11)/2 = 55
D_{10} = \$24,750(1/55) = \$450

(b) D.D.B.: BV_{10} = \$24,750$(1-0.2)^{10}$ = \$24,750(0.8)
= \$2,657.51
BV_9 = \$24750$(0.8)^9$ = \$3,321.89
D= \$3,321.89 - \$2,657.51 = \$664.38
(c) Since DDB method does not use the salvage value, the loss on disposal will be \$2,658.

11-4 (a) SOYD : SOYD = (40 x 41)/2 = 820
BV_{10} = \$100 + \$500 - (\$500/820)(820-465)
= \$383,537

BV_{20} = \$100 + \$500 - (\$500/820)(820-210)
= \$228,049

(b) D.D.B.: f = 2.000/40 = .05
BV_{10} = \$100 + \$500(1-0.05) = \$399,369

BV_{20} = \$100 + \$500(0.95) = \$279,244

11-5 P = \$26,000; L = \$2,000; n = 15; i = 10%
(a) S.L.: D_4 = (\$26,000-\$2,000)/15 = \$1,600

BV_4 = \$26,000 - \$1,600(4) = \$19,600

(b) SOYD: SOYD = (15 x 16)/2 = 120

D_4 = \$24,000(12/120) = \$2,400

BV_4 = \$26,000 - \$200(120 -66) = \$15,200

(c) D.D.B.: f = 2.000/15 = 0.13333

BV_4 = \$26,000$(1-0.13333)^4$ = \$14,671

BV_3 = \$26,000$(1-0.13333)^3$ = \$16,927

D_4 = \$16,927 - \$14,671 = \$2,256

(d) S.F.: BV_4 = \$26,000 - \$24,000(A/F,10%,15)(F/A,10%,4)
= \$22,495

BV_3 =\$26,000 - \$24,000(A/F,10%,15)(F/A,10%,3)
= \$23,500

D_4 = \$23,500 - \$22,495 = \$1,005

11-6 SOYD: SOYD = (5 x 6)/2 = 15

D_1 = \$35,000(5/15) = \$11,667

Profit = \$9,000 - (\$11,667 - \$7,000) = \$4,333

D.D.B.: f = 2.00/5 = 0.40

D_1 = \$35,000(0.40) = \$14,000

Profit = \$9,000 - (\$14,000 - \$7,000) = \$2,000

11-7 (a) (A/P,6%,5) = 0.23740

S.L. Dep. + Av. Int. = 1/5 + (0.06/2)(6/5) = 0.2360

Error = (0.23740 - 0.2360)/0.23740 = 0.59%

(b) (A/P,6%,10) = 0.13587

S.L. Dep. + Av. Int. = 1/10 +(0.06/2)(11/10) = 0.1330

Error = (0.13587 - 0.13300)/0.13587 = 2.11%

(c) (A/P,6%,20) = 0.08718

S.L. Dep. + Av. Int. = 1/20 +(0.06/2)(21/20) = 0.08150

Error = (0.08718 - 0.08150)/0.08718 = 6.52%

11-8 (a) (A/P,12%,5) = 0.27741

S.L.+Av. Int. = 1/5 + (.12/2)(6/5) = 0.27200

Error = (0.27741 - 0.27200)/0.27741 = 1.95%

(b) (A/P,12%,10) = 0.17698

S.L.+ Av. Int. = 1/10 + (.12/2)(11/10) = 0.16600

Error = (0.17698 - 0.16600)/0.17698 = 6.20%

11-8 continued

(c) (A/P,12%,20) = 0.13388

S.L.+ Av. Int. = 1/20 + (0.12/2)(21/20) = 0.11300

Error = (0.13388-0.11300)/0.13388 = 15.60%

11-9 For Vacuum Still:

NPW = 0 = -$110,000 + $38,000(P/A,i%,10) -$4,000(P/G,i%,10)

at 15% = -$110,000 + $38,000(5.019) -$4,000(16.9795) = $12,804

at 20% = -$110,000 + $38,000(4.192) -$4,000(12.8871) =- $2,252

i = 19.2% by interpolation

For Product Terminal:

NPW = 0 = -$110,000 + $5,000(P/A,i%,10) +$4,000(P/G,i%,10)

at 10% = -$110,000 + $5,000(6.144) +$4,000(22.8913) = +$12,285

at 12% = -$110,000 + $5,000(5.650) +$4,000(20.2541) = - $ 734

i = 11.9% by interpolation

11-10 For Vacuum Still, the "total profit" will be 10($38,000 - $11,000) - $4,000(9 x 10)/2 =$90,000; Average= $90,000/10= $9,000/yr.

Thus, by the "original book" method, i = $9,000/$110,000 = 8.2%

For Product Terminal, "total profit" = 10($5,000- $11,000) +$4,000(9 x 10)/2 = $120,000; Average = $120,000/10 =$12,000/yr. Thus, i = $12,000/$110,000 = 10.9%

11-11 Applying the "average book" method, and using the "annual profit" calculations in problem 11-10:

Vacuum Still: i = $9,000/$55,000= 16.4%

Product Terminal: i = $12,000/$55,000= 21.8%

11-12

Year	Start-of Year Book Value (Both)	Vacuum Still Annual Profit	Vacuum Still % RoR	Product Terminal Annual Profit	Product Terminal % RoR
1	$110	+$27	+24.5	-$6	-5.5
2	99	+23	+23.2	- 2	-2.0
3	88	+19	+21.6	+ 2	+2.3
4	77	+15	+19.5	+6	+7.8
5	66	+11	+16.7	+10	+15.2
6	55	+ 7	+12.7	+14	+25.5
7	44	+ 3	+ 6.8	+18	+40.9
8	33	- 1	- 3.0	+22	+66.7
9	22	- 5	-22.7	+26	+118.2
10	11	- 9	-81.8	+30	+272.7

11-12 Continued

Totals		+90	17.5	+120	+541.9
Average RoR			1.75%		54.2%

11-13 Year	Start-of year-Book	Vacuum Still Profit	RoR	Product Terminal Profit	RoR
1	$110	+$18	+16.4%	-$15	-13.6%
2	90	+ 16	+17.8	- 9	-10.0
3	72	+ 14	+19.4	- 3	- 4.2
4	56	+ 12	+21.4	+ 3	+ 5.4
5	42	+ 10	+23.8	+ 9	+21.4
6	30	+ 8	+26.7	+15	+50.0
7	20	+ 6	+30.0	+21	+105.0
8	12	+ 4	+33.3	+27	+225.0
9	6	+ 2	+33.3	+33	+550.0
10	2	0	0	+39	+1,950.0
Totals		+90	+221.1%	+120	+2,879.0
Average RoR			22.1%		287.9%

11-14 A correct method of analysis gives weight (through interest compounding) to the time of occurrence of the cash flows. In cases where the cash flows occur proportionately through time, the rankings will be the same as those arrived at by correct methods. However, in the case of the Vacuum Still, the largest portion of the cash flow occurred in the early years while that for the Product Terminal occurred in the later years.

11-15 The figures for rate of return on the Product Terminal illustrate even more graphically than those for the Vacuum Still a very important short coming of these approximate methods. That is, they depend very heavily on the amount of the investment capitalized on the company's books of account in determining the rate of return. Thus if an investment, such as a facility relayout, is expensed the first year, there is no capitalization and the rate of return goes to infinity. Correct discounting methods do not have this short coming since they consider cash flows, at the time of occurrence, rather than accounting figures. The students should see clearly after making the calculations of problem 11-13 that 195% of the 287.9% average rate of return for the Product Terminal comes from the year 10 rate of return calculation at which time the book value (denominator in the RoR calculation) is only $2,000. Clearly this is an absurd result.

11-16

Year	Net C F Before Deprecia.	(a) Straight Line Deprec.	(a) Straight Line Profit Before Income Tax	(b) Sum-of-Years-Digits Deprec.	(b) Sum-of-Years-Digits Profit Before Income Tax
1	+$16,000	$7,000	+$9,000	$11,667	$ 4,333
2	+ 19,000	9,000	10,000	12,667	6,333
3	+ 19,000	9,000	10,000	9,667	9,333
4	+ 17,000	9,000	8,000	6,667	10,333
	Totals	$34,000		$40,668	

(c) BV(S.L.) = $45,000 - $34,000 = +$11,000
BV(SOYD) = $45,000 - $40,668 = +$4,332

11-17

Year	Net C F Before Deprec.	(a) Straight Line Deprec.	(a) Straight Line Profit Before Income Tax	(b) Sum-of-Years-Digits Deprec.	(b) Sum-of-Years-Digits Profit Before Income Tax
1	+$16,000	$3,500	$12,500	$6,364	$9,636
2	+ 19,000	4,500	14,500	7,045	11,955
3	+ 19,000	4,500	14,500	6,727	12,273
4	+ 17,000	4,500	12,500	5,910	11,090
		(c) D.D.B			
1	+$16,000	$7,000	$ 9,000		
2	+ 19,000	7,600	11,400		
3	+ 19,000	6,080	12,920		
4	+ 17,000	4,864	12,136		

It is very clear that Equipment Rental would be better off by using either the SOYD or DDB methods, since it would have recovered more of its original investment and would have paid income taxes on smaller amounts each year.

11-18 (a) Find i so that
PW = 0 = -$45,000 +$9,000(P/A,i%,5) +$3,000(P/G,i%,5)
at i = 15%, PW = +$2,493
at i = 20%, PW = -$3,363; i = 17.13 by interpolation

PW = 0 = -$45,000 +$20,000(P/A,i%,5) - $3,000(P/G,i%,5)
at i = 20%, PW = +$102
at i = 25%, PW = -$3,831; i = 20.1%

(b) Model "A": Total "profit" = $9,000(5) +$3,000(4 x 5/2) -$9,000 = $30,000
Average profit = $30,000/5 = $6,000
i = $6,000/ $45,000 = 13.3%
Model "B": Total "profit" = $20,000(5) -$3,000(4 x 5/2) -$9,000(5) = $25,000
Average profit = $25,000/5 = $5,000
i = $5,000/$45,000 = 11.11%

11-18 Continued

(c) Model "A": i = $6,000/$22,500 = 26.67%
Model "B": i = $5,000/$22,500 = 22.22%

(d) Original Book and Average Book give no consideration to the timing of the "profits" received. Original Book would be a fairly good approximation if the salvage value were 100% of investment.

11-19 (Note well: The problem in the text refers to problem 11-8. That should be problem 11-18.)

	Model "A"			Model "B"		
Year	Start of Yr. Book Value	Profit After Deprec.	RoR %	Start of Yr. Book Value	Profit After Deprec	RoR %
1	$45,000	0	0	$45,000	+$11,000	24.44
2	36,000	+$3,000	8.33	36,000	+ 8,000	22.22
3	27,000	+ 6,000	22.22	27,000	+ 5,000	18.52
4	18,000	+ 9,000	50.00	18,000	+ 2,000	11.11
5	9,000	+12,000	133.33	9,000	- 1,000	-11.11
	Totals		213.88			65.18
Average RoR			42.78%			13.04%
True RoR			17.13%			20.1%

The results are severely distorted by this method since it does not take the timing and the magnitude of the "profits" into consideration.

11-20 (S) = Steel Building; (C) = Concrete Building

(a) EUAC(S) = -$150,000(A/P,15%,25) -$6,000 = -$29,205
EUAC(C) = -$200,000(A/P,15%,50) -$4,000 = -$34,028

(b) S.L. Dep.+ Av. Int. (S) =-$150,000/25
-$150,000(0.15/2)(26/25) -$6,000 = -$23,700
S.L. Dep.+Av. Int.(C) = -$200,000/50
-$200,000(0.15/2)(51/50 -$4,000 = - $23,300

(c) The long life for the concrete building increases the error in the approximation over that for the shorter life. The error is great enough to reverse the apparent monetary advantage of one building over the other.

(d) If ownership were compared with renting storage space, the S.L. Dep. + Av. Int Method understates the cost of both buildings and would lead to the wrong decision by indicating that the concrete building is the least expensive and much lower than the cost of renting.

11-21 (a) $D_{10} = (\$30{,}000 - \$4{,}000)/25 = \$1{,}040$

$BV_{10} = \$30{,}000 - \$1{,}040(10) = \$19{,}600$

(b) SOYD = $(25 \times 26/2) = 325$; $G = (\$30{,}000 - \$4{,}000)/325 = \$80$
$D_{10} = \$80(16) = \$1{,}280$

$BV_{10} = \$30{,}000 - \$80(325 - (15 \times 16/2)) = \$13{,}600$

(c) $f = 2.00/25 = 0.08$
$BV_9 = \$30{,}000(1 - 0.08)^9 = \$14{,}165$

$BV_{10} = \$30{,}000(1 - 0.08)^{10} = \$13{,}032$

$D_{10} = BV_9 - BV_{10} = \$1{,}133$

11-22 (a) $f = 1 - \sqrt[n]{L/P}$; $f = 1 - \sqrt[10]{0.02}$
$= 1 - 0.6762 = 0.3238$ or 32.38%

(b) $f = 1 - \sqrt[10]{0.05} = 25.89\%$

(c) $f = 1 - \sqrt[10]{0.20} = 14.87\%$

11-23 (a)

Year A	C F for Purchase B	C F for Make C	Net C F (C-B) D
0	0	-\$26,000	-\$26,000
1	-\$16,000	- 10,000	+ 6,000
2	- 18,500	- 12,000	+ 6,500
3	- 19,500	- 12,000	+ 7,500
4	- 14,500	- 9,000	+ 5,500
5	- 10,000	- 6,000	+ 4,000
6	- 6,000	- 3,000	+ 3,000
6	0	+ 2,000	+ 2,000
	- 84,500	- 76,000	+ 8,500

Find i so that NPW= $0 = -\$26{,}000 + \$6{,}000(P/F,i\%,1) + \$6{,}500(P/F,i\%,2) + \$7{,}500(P/F,i\%,3) + \$5{,}500(P/F,i\%,4) + \$4{,}000(P/F,i\%,5) + \$5{,}000(P/F,i\%,6)$

$i = 9.34\%$ by interpolation between 9% and 10%.

(b)

Yr.	C F Before Taxes	Depreciation	Taxable Income	C F For Taxes	C F After Income Tax
0	-$26,000				-$26,000
1	+ 6,000	-$4,000	+$2,000	-$1,000	+ 5,000
2	+ 6,500	- 4,000	+ 2,500	- 1,250	+ 5,250
3	+ 7,500	- 4,000	+ 3,500	- 1,750	+ 5,750
4	+ 5,500	- 4,000	+ 1,500	- 750	+ 4,750
5	+ 4,000	- 4,000	+ 0	- 0	+ 4,000
6	+ 3,000	- 4,000	- 1,000	+ 500	+ 3,500
6	+ 2,000				+ 2,000

Find i so that NPW = 0 = -$26,000 + $5,000(P/F,i%,1) +$5,250(P/F,i%,2) +$5,750(P/F,i%,3) +$4,750(P/F,i%,4) +$4,00(P/F,i%,5) +$5,500(P/F,i%,6)

i = 4.56% after income taxes, by interpolation between 4.5% and 5%.

11-24 (a) S.L.: BV_{10} = $700,000 - $600,000(10/20) = $400,000

(b) SOYD: BV_{10} = $700,000 -($600,000/210)(210 -(10 x 11/2)) = $257,143

(c) D.D.B.: f = 2.000/20 = 0.10
BV_{10} = $100,000 + $600,000(1 - 0.10) = $309,207

(d) Find i so that NPW = 0 = -$700,000 +$120,000(P/A,i%,10) +$500,000(P/F,i%,10)
i = 15.79% by interpolation between 15% and 16%.

11-25 On a graph plot the error vs the years for the data in the following table:

Years	i = 10 % (A/P,10%,n)	S.L.dep. +Av.Int.	Error	i = 15% (A/P,15%,n)	S.L. Dep. +Av. Int.	Error
5	0.26380	0.26000	1.4%	0.29832	0.29000	2.8%
10	0.16275	0.15500	4.8%	0.19925	0.18250	8.4%
15	0.13147	0.12000	8.7%	0.17102	0.14667	14.2%
20	0.11746	0.10250	12.7%	0.15976	0.12875	19.4%
25	0.11017	0.09200	16.5%	0.15470	0.11800	23.7%
30	0.10608	0.08500	19.9%	0.15230	0.11083	27.2%

This exercise should demonstrate the rapidity with which the error resulting from using the straight line depreciation plus average interest increases as the interest rate and/or the number of years increases.

11-26 (a) S.L.: D = $150,000/15 = $10,000
BV = $200,000 - $10,000(10) = $100,000

(b) SOYD: SOYD = (15 x 16/2) = 120; G =$150,000/120 = $1,250
D = $1,250(6) = $7,500
BV = $200,000 - ($150,000 - $1,250(5 x 6/2)) = $68,750

11-26 Continued

(c) D.D. B.: $f = 2.000/15 = 0.13333$

$BV(9) = \$50,000 + \$150,000(1 - 0.13333)^9 = \$91,377$

$BV(10) = \$50,000 + \$150,000(1 - 0.13333)^{10} = \$85,860$

$D = \$91,377 - \$85,860 = \$5,517$

(d) S.F.: $i = 10\%$

$BV(9) = \$200,000 - \$150,000(A/F,10\%,15)(F/A,10\%,9)$
$= \$135,900$

$BV(10) = \$200,000 - \$150,000(A/F,10\%,15)(F/A,10\%,10)$
$= \$124,769$

$D = \$135,900 - \$124,769 = \$11,131$

11-27 (a) S.L.: $D = (\$25,000 - \$4,000)/6 = \$3,500$

$BV = \$25,000 - \$3,500(3) = \$14,500$

(b) SOYD: $SOYD = (6 \times 7/2) = 21$; $G = \$21,000/21 = \$1,000$

$D = \$1,000(4) = \$4,000$

$BV = \$25,000 - \$1,000(21 - (3 \times 4/2)) = \$10,000$

(c) D.D.B.: $f = 2.000/6 = 0.33333$

$BV(2) = \$25,000(1 - 0.33333)^2 = \$11,111$

$BV(3) = \$25,000(1 - 0.33333)^3 = \$7,408$

$D = \$11,111 - \$7,408 = \$3,703$

(d) S.F.: $i = 6\%$

$BV(2) = \$25,000 - \$21,000(A/F,6\%,6)(F/A,6\%,2) = \$18,798$

$BV(3) = \$25,000 - \$21,000(A/F,6\%,6)(F/A,6\%,3) = \$15,414$

$D = \$18,798 - \$15,414 = \$3,384$

11-28 (a) S.L.: $D(1) = \$10,000(0.20) + (\$23,000 - \$4,000)/6 = \$5,166$

$D(2) = (\$23,000 - \$4,000)/6 = \$3,166$

(b) SOYD: $SOYD = (6 \times 7/2) = 21$

$D(1) = \$2,000 + \$19,000(6/21) = \$7,429$

$D(2) = (\$19,000)(5/21) = \$4,524$

(c) D.D.B.: $f = 2.000/6 = 0.33333$

$D(1) = \$2,000 + \$23,000(0.33333) = \$9,667$

$D(2) = (\$23,000)(1 - 0.33333)(0.33333) = \$5,111$

CHAPTER 12

Estimating Income-Tax Consequences of Certain Decisions

12-1

Year	Cash Flow Before Income Taxes	Write-Off of Initial Outlay for Tax Purposes	Influence on Taxable Income	Influence of Income Taxes on Cash Flow	Cash Flow After Income Taxes
0	-$78,000				-$78,000
1	+ 16,300	-$12,000	+$4,300	-$2,064	+ 14,236
2	+ 16,300	- 11,000	+ 5,300	- 2,544	+ 13,756
3	+ 16,300	- 10,000	+ 6,300	- 3,024	+ 13,276
4	+ 16,300	- 9,000	+ 7,300	- 3,504	+ 12,796
.	.	.	.	.	.
.	.	.	.	.	.
12	+ 16,300	- 1,000	+15,300	- 7,344	+ 8,956

Rate of return before income taxes:

$0 = -\$78,000 + \$16,300(P/A,i\%,12) \quad i = 18.0\%$

Rate of return after income taxes:

$0 = -\$78,000 + \$14,236(P/A,i\%,12)-\$480(P/G,i\%,12) \quad i = 11.3\%$

12-2

(Headings same as Problem 12-1)

0	-$78,000				-$78,000
1-12	+ 16,300	-$ 6,500	+$9,800	-$4,704	+ 11,596

Rate of return after income taxes:

$0 = -\$78,000 + \$11,596(P/A,i\%,12) \quad i = 10.3\%$

12-3

This is same as Problem 12-1 except for year 0, which is

$\$-\$78,000 + \$7,800 = -\$70,200$

Rate of return after income taxes:

$0 = -\$70,200+\$14,236(P/A,i\%,12)-\$480(P/G,i\%,12) \quad i = 13.8\%$

12-4

(Headings same as Problem 12-1)

0	-$78,000			+$7,800	-$70,200
1-5	+ 16,300	-$15,600	+$ 700	- 336	+ 15,964
6-12	+ 16,300	0	+ 16,300	- 7,824	+ 8,476

Rate of return after income taxes:

0 = -$70,200 + $15,964(P/A,i%,5)
+ $8,476(P/A,i%,12 - P/A,i%,5)   i = 15.3%

12-5

(Headings same as Problem 12-1)

0	-$78,000				-$78,000
1-12	+ 16,300	-$ 3,900	+$12,400	-$5,952	+ 10,348
13-20	0	- 3,900	- 3,900	+ 1,872	+ 1,872

Rate of return after income taxes:

0 = -$78,000 + $10,346(P/A,i%,12)
+ $1,872(P/A,i%,20 - P/A,i%,12)   i = 9.0%

12-6

(Headings same as 12-1)

Book value after 12 years = $78,000 - 12($3,900) = $31,200 This will be the "loss of disposal" if machinery is disposed of with zero net salvage.

0	-$78,000				-$78,000
1-12	+ 16,300	-$ 3,900	+$12,400	-$ 5,952	+ 10,348
12		- 31,200	- 31,200	+ 14,976	+ 14,976

Rate of return after income taxes:

0 = -$78,000+$10,348(P/A,i%,12) + $14,976(P/F,i%,12)
i = 9.3%

12-7

(Headings same as Problem 12-1)

0	-$126,000				-$126,000
1	+ 34,000	-$28,000	+$ 6,000	-$ 2,760	+ 31,240
2	+ 34,000	- 24,500	+ 9,500	- 4,370	+ 29,630
3	+ 34,000	- 21,000	+ 13,000	- 5,980	+ 28,020
.	.	.	.	.	.
8	+ 34,000	- 3,500	+ 30,500	- 14,030	+ 19,970

12-7 continued

Rate of return before income taxes:

$0 = -\$126{,}000 + \$34{,}000(P/A,i\%,8)$ $i = 21.3\%$

Rate of return after income taxes:

$0 = -\$126{,}000 + \$31{,}240(P/A,i\%,8) - \$1{,}610(P/G,i\%,8)$
$i = 13.5\%$

12-8

(Headings same as Problem 12-1)

0	-$126,000				-$126,000
1-8	+ 34,000	-$15,750	+18,250	-$8,395	+ 25,605

Rate of return after income taxes:
$0 = -\$126{,}000 + \$25{,}605(P/A,i\%,8)$ $i = 12.3\%$

12-9

This is the same as Problem 12-7 except for year 0, which is -$126,000 + $12,600 = -$113,400

Rate of return after income taxes:
$0 = -\$113{,}400+\$31{,}240(P/A,i\%,8)-\$1{,}610(P/G,i\%,8)$ $i = 17.0\%$

12-10

(Headings same as Problem 12-1)

0	-$126,000				-$126,000
1	+ 34,000	-$36,000	-$ 2,000	+$ 920	+ 34,920
2	+ 34,000	- 30,000	+ 4,000	- 1,840	+ 32,160
3	+ 34,000	- 24,000	+ 10,000	- 4,600	+ 29,400
4	+ 34,000	- 18,000	+ 16,000	- 7,360	+ 26,640
5	+ 34,000	- 12,000	+ 22,000	-10,120	+ 23,880
6	+ 34,000	- 6,000	+ 28,000	-12,880	+ 21,120
7	+ 34,000		+ 34,000	-15,640	+ 18,360
8	+ 34,000		+ 34,000	-15,640	+ 18,360

Rate of return after income taxes:
$0 = -\$126{,}000+\$34{,}920(P/A,i\%,7) - \$2{,}760(P/G,i\%,7)$
$+ \$18{,}360(P/F,i\%,8)$ $i = 14.2\%$

12-11

This is the same as Problem 12-10 except for year 0, which is -$126,000 + $12,600 = -$113,400

Rate of return after income taxes:
$0 = -\$113{,}400 + \$34{,}920(P/A,i\%,7) - \$2{,}760(P/G,i\%,7)$
$+ \$18{,}360(P/F,i\%,8)$ $i = 18.0\%$

12-12

(Headings same as Problem 12-1)

0	-$126,000				-$126,000
1-8	+ 34,000	-$6,300	+$27,700	-$12,742	+ 21,258
9-20		- 6,300	- 6,300	+ 2,898	+ 2,898

Rate of return after income taxes:

$0 = -\$126,000 + \$21,258(P/A,i\%,8)$
$+\$2,898(P/A,i\%,20 - P/A,i\%,8)$ $i = 9.4\%$

12-13

(Headings same as Problem 12-1)

0	-$110,000	-$110,000	-$110,000	+$66,000	-$44,000
1-10	+ 26,300		+ 26,300	- 12,887	+ 13,413

Rate of return after income taxes:

$0 = -\$44,000 + \$13,413(P/A,i\%,10)$ $i = 28.0\%$

12-14

Consider a $20,000 asset; the example used in the problem statement. Solution requires evaluating the difference in tax saving under the two proposals, the 7% ITC and the 10% ITC. The analysis following assumes tax rates of 30% and 50%, lives of 10 and 20 years, and SL and SYD Depreciation.

10-yr. life, SL Dep. - Tax Savings

	7% ITC		10% ITC		Difference 7%-10%	
Yr	30%TR	50%TR	30%TR	50%TR	30%TR	50%TR
0	$420	$700	$600	$1,000	-$180	-$300
1-10	600	1,000	540	900	+ 60	+ 100
Rate of Return using 7% ITC rather than 10%					31.1%	31.1%

20-yr. life, SL Dep. - Tax Savings

	7% ITC		10% ITC		Difference 7%-10%	
Yr	30%TR	50%TR	30%TR	50%TR	30%TR	50%TR
0	$420	$700	$600	$1,000	-$180	-$300
1-20	300	500	270	450	+ 30	+ 50
Rate of Return using 7% ITC rather than 10%					15.8%	15.8%

12-14 continued

10-yr. life, SYD Dep. - Tax Savings

Yr	7% ITC 30%TR	7% ITC 50%TR	10% ITC 30%TR	10% ITC 50%TR	Difference 7%-10% 30%TR	Difference 7%-10% 50%TR
0	$420	$700	$600	$1,000	-$180	-$300
1	1,091	1,818	982	1,636	+ 109	+ 182
2	982	1,636	884	1,472	+ 98	+ 164
3	873	1,455	786	1,308	+ 87	+ 146
.	.	.	.	.	.	.
.	.	.	.	.	.	.
10	109	182	98	164	+ 11	+ 18
Rate of Return using 7% ITC rather than 10%					48.5%	48.5%

In general, the 7% ITC without reduction in the depreciation base is preferable to the 10% ITC with a corresponding 10% reduction in the depreciation base. More liberal depreciation methods enhance this preference considerably. A preference for the 10% ITC as described may occur for extremely long lived assets and particularly if SL Depreciation is required. Buildings would fit in this category. This results from the fact that the break-even rate of return decreases as depreciable life increases. This can be seen from the rate of return difference calculations using SL Dep. The incremental tax rate is not relevant to the decision. Extremely high interest rates would tend to favor the 10% ITC as described.

12-15

Income tax = \$9,250 + 0.30(\$60,000 - \$50,000) = \$12,250

The applicable tax rate for economy studies should be the rate on the highest increment of income that is subject to tax. In this case the rate is 30%.

12-16

The combined incremental rate is s + (1 - s)f. In all four cases the state tax is not enough to change the applicable federal incremental rate.

For \$30,000, combined rate is .06 + (.94)(.20) = .248 or 24.8%
For \$60,000, " " " .06 + (.94)(.30) = .342 or 34.2%
For \$90,000, " " " .06 + (.94)(.40) = .436 or 43.6%
For \$120,000, " " " .06 + (.94)(.46) = .4924 or 49.24%

12-17

The incremental rates involved are 34% if adjusted gross income is between \$22,000 and \$24,500, 39% between \$24,500 and \$29,500, and 44% between \$29,800 and \$32,000. Clearly, the average incremental rate is about 39%. For simplicity, this might be rounded off to 40%.

12-18

(a) $0 = -\$50{,}000 + (\$16{,}000 - \$3{,}000)(P/A,i\%,5) + \$15{,}000(P/F,i\%,5)$ $\quad i = 15.8\%$

(b) $0 = -\$50{,}000 + \$16{,}000(P/A,i\%,5) + \$15{,}000(P/F,i\%,5)$ $\quad i = 23.3\%$

12-19

Investment tax credit at zero date will be 6-2/3% of \$35,000, which is \$2,333. Estimated extra annual income tax 0.40(\$16,000 - \$3,000 - \$7,000) = \$2,400

To compute prospective after-tax rate of return:

$0 = -\$50{,}000 + \$2{,}333 + (\$16{,}000 - \$3{,}000 - \$2{,}400)(P/A,i\%,5) + \$15{,}000(P/F,i\%,5)$ $\quad i = 11.3\%$

12-20

(Headings are same as Problem 12-1)

0	-\$50,000			+\$2,333	-\$47,667
1	+ 13,000	-\$11,667	+\$ 1,333	- 533	+ 12,467
2	+ 13,000	- 9,333	+ 3,667	- 1,467	+ 11,533
3	+ 13,000	- 7,000	+ 6,000	- 2,400	+ 10,600
4	+ 13,000	- 4,667	+ 8,333	- 3,333	+ 9,667
5	+ 13,000	- 2,333	+ 10,667	- 4,267	+ 8,733
5	+ 15,000				+ 15,000

Rate of return after income taxes:

$0 = -\$47{,}667 + \$12{,}467(P/A,i\%,5) - \$933(P/G,i\%,5) + \$15{,}000(P/F,i\%,5)$ $\quad i = 11.9\%$

The use of years-digits depreciation would increase the prospective rate of return from 11.3% to 11.9%.

12-20 continued

With years-digits depreciation based on a 4-year life and with an investment tax credit of $1,167 (3-1/3% of $35,000), the table of prospective cash flows becomes:

0	-$50,000			$1,167	-$48,833
1	+ 13,000	-$14,000	-$ 1,000	+ 400	+ 13,400
2	+ 13,000	- 10,500	+ 2,500	- 1,000	+ 12,000
3	+ 13,000	- 7,000	+ 6,000	- 2,400	+ 10,600
4	+ 13,000	- 3,500	+ 9,500	- 3,800	+ 9,200
5	+ 13,000		+ 13,000	- 5,200	+ 7,800
5	+ 15,000				+ 15,000

Rate of return after income taxes:

$$0 = -\$48{,}833 + \$13{,}400(P/A,i\%,5) - \$1{,}400(P/G,i\%,5) + \$15{,}000(P/F,i\%,5) \qquad i = 11.3\%$$

The combined effect of the more rapid depreciation and the reduced investment tax credit is to reduce the prospective rate of return from 11.9% to 11.3%.

A more sophisticated analysis involves examining the differences in after-tax cash flows in the two tabulations. The choice of faster depreciation and lower investment tax credit causes the following differences in cash flow:

0	-$1,334
1	+ 933
2	+ 467
3	0
4	- 467
5	- 933

Although the foregoing is a cash flow series that has two reversals of sign and therefore is a series of the type discussed in Appendix B, a mere inspection of the series shows that it is unattractive to sacrifice part of the investment tax credit by shortening the life used for tax purposes. Even if there were no negative cash flows in years 4 and 5, the prospective rate of return would be very low.

12-21

Annual cost of rental = $25,000
Annual cost of ownership = $3,000 + $110,000(0.12) + $10,780
= $26,980

12-22

Annual cost with proposal rejected = \$32,000
Annual cost with proposal accepted = \$5,700 + \$110,000(A/P,12%,10) + \$3,087 + \$980(A/G,12%,10) = \$31,768

12-23

Annual cost with proposal rejected = \$32,000
Annual cost with proposal accepted = \$5,700 + \$110,000(A/P,12%,10) + \$7,497 = \$32,665

12-24

A special difficulty is caused because the tax consequences extend beyond the 10-year study period that seems appropriate for a comparison of annual costs. The positive cash flow from tax savings of \$2,156 a year for years 11-25 might be converted to an equivalent uniform annual figure for years 1-10. A possible method for this comparison is to calculate present worth at zero date as \$2,156(P/A,12%,25 - P/A,12%,10) = \$4,278. This converts to a uniform figure for years 1-10 of \$4,278(A/P,12%,10) = \$757.

Annual cost for years 1-10 with proposal rejected = \$36,300
Equivalent uniform annual cost for years 1-10 with proposal accepted = \$10,000 + \$110,000(A/P,12%,10) + (\$10,731 - \$757) = \$39,442

12-25

Annual cost with proposal rejected = \$32,000
Annual cost with proposal accepted = \$5,700 + \$110,000(A/P,12%,10) + \$2,107 + (\$12,807 - \$2,107)(P/A,12%,10 - P/A,12%,5) (A/P,12%,10) = \$31,148

12-26

Annual cost with proposal rejected = \$26,300
Annual cost with proposal accepted = (\$110,000 - \$53,900)(A/P,12%,10) + \$12,887 = \$22,816

12-27

PW of rental = \$25,000(P/A,11%,10) = \$147,220
PW of ownership = \$110,000 - \$110,000(P/F,11%,10) + \$3,000(P/A,11%,10) + \$10,780(P/A,11%,10) = \$152,410

12-28

PW with proposal rejected = \$32,000(P/A,11%,10) = \$188,450
PW with proposal accepted = \$110,000 + \$5,700(P/A,11%,10)
+ \$3,067(P/A,11%,10) + \$980(P/G,11%,10) = \$182,720

12-29

PW with proposal rejected = \$32,000(P/A,11%,10) = \$188,450
PW with proposal accepted = \$110,000 + \$5,700(P/A,11%,10)
+ \$7,497(P/A,11%,10) = \$187,720

12-30

PW with proposal rejected = \$36,300(P/A,11%,10) = \$213,770
PW with proposal accepted = \$110,000 + \$10,000(P/A,11%,10)
+ \$10,731(P/A,11%,10) - \$2,136(P/A,11%,25 - P/A,11%,10)
= \$226,670

12-31

PW with proposal rejected = \$32,000(P/A,11%,10) = \$188,450
PW with proposal accepted = \$110,000 + \$5,700(P/A,11%,10)
+ \$2,107(P/A,11%,10) + (\$12,807 - \$2,107)(P/A,11%,10
- P/A,11%,5) = \$179,450

12-32

PW with proposal rejected = \$26,300(P/A,11%,10) = \$154,880
PW with proposal accepted = \$110,000 - \$53,900 +
\$12,887(P/A,11%,10) = \$131,990

12-33

(Headings same as Problem 12-1)
With tax rate of 25% and no investment tax credit:

0	-\$110,000				-\$110,000
1	+ 26,300	-\$20,000	+\$ 6,300	-\$1,575	+ 24,725
2	+ 26,300	- 18,000	+ 8,300	- 2,075	+ 24,225
	.	.	.	.	.
10	+ 26,300	- 2,000	+ 24,300	- 6,075	+ 20,225

0 = -\$110,000 + \$24,725(P/A,i%,10) - \$500(P/G,i%,10)   i = 16.4%

With a 10% investment tax credit the cash flow is the same for years 1 through 10. For year 0 there is an \$11,000 positive cash flow caused by the tax credit; the net cash flow for year 0 is -\$99,000.

0 = -\$99,000 + \$24,725(P/A,i%,10) - \$500(P/G,i%,10)   i = 19.5%

12-34

(Headings same as Problem 12-1)
With tax rate of 25% and no investment tax credit:

0	-\$78,000				-\$78,000
1	+ 16,300	-\$12,000	+\$ 4,300	-\$1,075	+ 15,225
2	+ 16,300	- 11,000	+ 5,300	- 1,325	+ 14,975
.	.	.	.	.	.
12	+ 16,300	- 1,000	+15,300	- 3,825	+ 12,475

0 = -\$78,000 + \$15,225(P/A,i%,12) - \$250(P/G,i%,12)   i = 14.8%

With a 10% investment tax credit the cash flow is the same for years 1 through 12. For year 0 there is a \$7,800 positive cash flow caused by the tax credit; the net cash flow for year 0 is -\$70,200.

0 = -\$70,200 + \$15,225(P/A,i%,12) - \$250(P/G,i%,12)   i = 17.4%

12-35

(Headings same as Problem 12-1)

(a) Rate of return before income taxes is 21.3%, already calculated in Problem 12-7.

(b) With tax rate of 30% and no investment tax credit:

0	-\$126,000				-\$126,000
1	+ 34,000	-\$28,000	+\$ 6,000	-\$1,800	+ 32,200
2	+ 34,000	- 24,500	+ 9,500	- 2,850	+ 31,150
.	.	.	.	.	.
8	+ 34,000	- 3,500	+ 30,500	- 9,150	+ 24,850

0 = -\$126,000 + \$32,200(P/A,i%,8) - \$1,050(P/G,i%,8)   i = 16.4%

(c) With a 10% investment tax credit the cash flow is the same for years 1 through 8. For year 0 there is a \$12,600 positive cash flow caused by the tax credit; the net cash flow for year 0 is -\$113,400

0 = -\$113,400 + \$32,200(P/A,i%,8) - \$1,050(P/G,i%,8)   i = 19.9%

12-36

(a) Before-tax rate of return is 21.3% as before
(b) Table is same as in part (b) of the preceding problem except that taxes in the next-to-last column are -\$1,200, -\$1,900, ... -\$6,400, and after-tax cash flows in final column are -\$126,000 + \$32,800, +\$32,100, ... +\$27,900.

0 = -\$126,000 + \$32,800(P/A,i%,8) - \$700(P/G,i%,8)   i = 18.0%

12-36 continued

(c) Net cash flow for year 0 is -\$113,400 as in Problem 12-35(c

0 = -\$113,400 + \$32,800(P/A,i%,8) - \$700(P/G,i%,8) i = 21.8%

12-37

(Headings are same as in Problem 12-1)

(a) 0 = -\$63,000 + \$16,300(P/A,i%,15) i = 25.0%

(b) With tax rate of 48%, straight-line depreciation based on 15-year life, and no investment tax credit:

0	-\$63,000				-\$63,000
1-15	+ 16,300	-\$4,200	+\$12,100	-\$5,808	+ 10,492

0 = \$63,000 + \$10,492(P/A,i%,15) i = 14.5%

(c) With tax rate of 48%, years-digits depreciation based on 15-year life, and no investment tax credit:

0	-\$63,000				-\$63,000
1	+ 16,300	-\$7,875	+\$8,425	-\$4,044	+ 12,256
2	+ 16,300	- 7,350	+ 8,950	- 4,296	+ 12,004
.	.	.	.	.	.
15	+ 16,300	- 525	+15,775	- 7,572	+ 8,728

0 = -\$63,000 + \$12,256(P/A,i%,15) - \$252(P/G,i%,15) i = 15.7%

(d) With tax rate of 48%, years-digits depreciation based on 9-year life, and 10% investment tax credit.

0	-\$63,000			+\$6,300	-\$56,700
1	+ 16,300	-\$12,600	+\$ 3,700	- 1,776	+ 14,524
2	+ 16,300	- 11,200	+ 5,100	- 2,448	+ 13,852
.	.	.	.	.	.
9	+ 16,300	- 1,400	+ 14,900	- 7,152	+ 9,148
10	+ 16,300	0	+ 16,300	- 7,824	+ 8,476
11	+ 16,300	0	+ 16,300	- 7,824	+ 8,476
.	.	.	.	.	.
15	+ 16,300	0	+ 16,300	- 7,824	+ 8,476

0 = -\$56,700 + \$14,524(P/A,i%,10) - \$672(P/G,i%,10)
+ \$8,476(P/A,i%,15 - P/A,i%,10) i = 19.8%

12-38

(a) Before-tax i = 25.0% just as before

(b) Annual income tax = 0.40($12,100) = $4,840
Annual positive cash flow years 1-15 = $16,300 - $4,840 = $11,460

0 = -$63,000 + $11,460(P/A,i%,15) i = 16.3%

(c) Tax in year 1 = $8,425(0.40) = $3,370. Each year this increases by 0.40($525) = $210.

0 = -$63,000 + $12,930(P/A,i%,15) - $210(P/G,i%,15) i = 17.4%

(d) Table from Problem 12-37(d) is modified so that cash flow for year 1 is $16,300 - 0.40($3,700) = +$14,820. Gradient for years 1-10 is 0.40($1,400) = $560. Cash flow years 11-15 is $9,780.

0 = -$56,700 + $14,820(P/A,i%,10) - $560(P/G,I%,10)
+ $9,780(P/A,i%,15 - P/A,i%,10) i = 21.5%

12-39

(a) Before-tax i = 25.0% just as before.

(b) Annual income tax = 0.32($12,100) = $3,872
Annual positive cash flow years 1-15 = $16,300 - $3,872 = $12,428

0 = -$63,000 + $12,428(P/A,i%,15) i = 18.1%

(c) Tax in year 1 = $8,425(0.32) = $2,696. Each year this increases by 0.32($525) = $168.

0 = -$63,000 + $13,604(P/A,i%,15) - $168(P/G,i%,15) i = 19.0%

(d) Table from Problem 12-37(d) is modified so that cash flow for year 1 is $16,300 - 0.32($3,700) = +$15,116. Gradient for years 1-10 is 0.32($1,400) = $448. Cash flow years 11-15 is $11,084.

0 = -$56,700 + $15,116(P/A,i%,10) - $448(P/G,i%,10)
+ $11,084(P/A,i%,15 - P/A,i%,10) i = 23.0%

12-40

With the investment tax credit at year 1:

0 = -$110,000 + 23,213(P/A,i%,10) - $980(P/G,i%,10)
+ $11,000(P/F,i%,1) i = 14.8%

With the investment tax credit at year 3:

0 = -$110,000 + $23,213(P/A,i%,10) - $980(P/G,i%,10)
+ $11,000(P/F,i%,3) i = 14.3%

12-41

Taxable incomes are the same as in Tables 12-2 and 12-3. With an applicable tax rate of 32% for years 1, 2, and 3, a rate of 42% for years 4 and 5, and a rate of 49% for years 6 through 10, estimated cash flows after taxes with the two depreciation methods will be as follows:

Year	Years-digits method	Straight-line method	Difference (YD - SL)
0	-$110,000	-$110,000	0
1	+ 24,284	+ 21,404	+$2,880
2	+ 23,644	+ 21,404	+ 2,240
3	+ 23,004	+ 21,404	+ 1,600
4	+ 21,134	+ 19,874	+ 1,260
5	+ 20,294	+ 19,874	+ 420
6	+ 18,313	+ 18,803	- 490
7	+ 17,333	+ 18,803	- 1,470
8	+ 16,353	+ 18,803	- 2,450
9	+ 15,373	+ 18,803	- 3,430
10	+ 14,393	+ 18,803	- 4,410

Trial-and-error present worth calculations indicate a prospective rate of return of about 13.4% if years-digits depreciation is used for tax purposes and about 12.9% if straight-line is used. Nevertheless, such calculations are not necessary to guide this decision. The right-hand column of the foregoing table is the one that should be analyzed. It shows the differences in cash flow to be caused by the choice between the two depreciation methods. (Note that the figures in this column really are independent of the merits of the investment proposal; they depend only on the applicable tax rates and the depreciation deductions from taxable income.)

Whenever the applicable tax rate is not expected to change during a study period and the taxpayer is allowed a choice among different depreciation methods, it is desirable to select the method that writes off the cost of fixed assets as rapidly as possible. If the total income tax caused by an investment in fixed assets will not be changed by the choice of a depreciaiton method, it is better to pay the tax later rather than sooner.

When the applicable tax rate is expected to rise, as in this problem, the choice of the more rapid depreciation method (years-digits in this case) leads to less taxes in the immediate future but more taxes in the long run. In this case a total of $8,400 less taxes for years 1-5 will cause $12,250 more taxes for years 6-10. Trial-and-error present worth calculations show that the present worth of the series in the right-hand column of the foregoing table

12-41 continued

is zero at an interest rate of about 6%. In the language of Appendix B the choice between the two depreciation methods is a borrowing-financing decision. If it is attractive to secure funds at an after-tax cost of 6%, years-digits depreciation should be selected. If 6% is viewed as too high, all things considered, straight-line depreciation should be used.

12-42

Gross income as a tax base disregards the expenses necessary to produce the income. Therefore it cannot reflect capacity to make tax payments. Moreover, a tax on gross income is socially undesirable because it gives an unfair competitive advantage to those integrated manufacturing enterprises that start with producing raw materials and end with marketing a finished product.

12-43

Problem	Payoff period
12-1	6.0 years
12-2	6.7 years
12-3	5.3 years
12-4	4.4 years
12-5	7.5 years
12-6	7.5 years

12-44

Problem	Payoff period
12-7	4.4 years
12-8	4.9 years
12-9	3.9 years
12-10	4.1 years
12-11	3.6 years
12-12	5.9 years

12-45

The first $1,800 will be taxed at 39% and the remaining $4,200 at 44%. Added tax will be $702 + $1,848 = $2,550. She will have $3,450 of the $6,000 left after paying her extra federal income taxes.

12-46

For each extra dollar of corporate income 46¢ is paid in income taxes. If the remaining 54¢ is paid as taxable dividends, 0.55(54¢) = 29.7¢ will be paid as personal income taxes. Thus the original extra dollar of corporate income will cause 46¢ + 29.7¢ = 75.7¢ of income taxes.

12-47

(Headings are same as in Problem 12-1.)

0	-$110,000			+$11,000	-$99,000
1	+ 26,300	-$20,000	+$ 6,300	- 4,769	+ 21,531
2	+ 26,300	- 18,000	+ 8,300	- 6,283	+ 20,017
.	.	.	.	.	.
10	+ 26,300	- 2,000	+ 24,300	- 18,395	+ 7,905

0 = -$99,000 + $21,531(P/A,i%,10) - $1,514(P/G,i%,10)   i = 9.6%

12-48

(Ex. 12-1) Annual income taxes are 0.30($22,000) = $6,600. After-tax cash flow years 1-10 = $22,000 - $6,600 = $15,400. Rate of return = $15,400 ÷ $110,000 = 0.140 or 14.0%

(Ex. 12-2) Income tax is 0.30($6,300) = $1,890 in year 1. Taxes increase by 0.30($2,000) = $600 a year.

0 = -$110,000 + ($26,300 - $1,890)(P/A,i%,10)
- $600(P/G,i%,10)   i = 15.6%

(Ex. 12-3) Annual income taxes are 0.30($15,300) = $4,590
0 = -$110,000 + ($26,300 - $4,590)(P/A,i%,10)   i = 14.8%

(Ex. 12-4) Annual income taxes for years 1-10 are 0.30($21,900) = $6,570. Reduction in annual income taxes for years 11-25 is 0.30($4,400) = $1,320.

0 = -$110,000 + ($26,300 - $6,570)(P/A,i%,10)
+ $1,320(P/A,i%,25 - P/A,i%,10)   i = 12.9%

(Ex. 12-5) Annual income taxes for years 1-5 are 0.30($4,300) = $1,290. For years 6-10 they are 0.30($26,300) = $7,890.

0 = -$110,000 + ($26,300 - $1,290)(P/A,i%,5)
+ ($26,300 - $7,890)(P/A,i%,10 - P/A,i%,5)   i = 16.2%

(Ex. 12-6) Reduction in income tax for year 0 is 0.30($110,000) = $33,000. Increase in annual income taxes for years 1-10 = 0.30($26,300) = $7,890

0 = -$110,000 + $33,000 + ($26,300 - $7,890)(P/A,i%,10)
i =20.1%

12-49

(Ex. 12-1) Annual income taxes are 0.15($22,000) = $3,300. After-tax cash flow years 1-10 = $22,000 - $3,300 = $18,700. Rate of return = $18,700 ÷ $110,000 = 0.170 or 17.0%

(Ex. 12-2) Income tax is 0.15($6,300) = $945 in year 1. Taxes increase by 0.15($2,000) = $300 a year.

$$0 = -\$110,000 + (\$26,300 - \$945)(P/A,i\%,10) - \$300(P/G,i\%,10) \qquad i = 17.9\%$$

(Ex. 12-3) Annual income taxes are 0.15($15,300) = $2,295

$$0 = -\$110,000 + (\$26,300 - \$2,295)(P/A,i\%,10) \qquad i = 17.5\%$$

(Ex. 12-4) Annual income taxes for years 1-10 are 0.15($21,900) = $3,285. Reduction in annual income taxes for years 11-25 is 0.15($4,400) = $660.

$$0 = -\$110,000 + (\$26,300 - \$3,285)(P/A,i\%,10) + \$660(P/A,i\%,25 - P/A,i\%,10) \qquad i = 16.5\%$$

(Ex. 12-5) Annual income taxes for years 1-5 are 0.15($4,300) = $645. For years 6-10 they are 0.15($26,300) = $3,945.

$$0 = -\$110,000 + (\$26,300 - \$645)(P/A,i\%,5) + (\$26,300 - \$3,945)(P/A,i\%,10 - P/A,i\%,5) \qquad i = 18.2\%$$

(Ex. 12-6) Reduction in income taxes for year 0 is 0.15($110,000) = $16,500. Increase in annual income taxes for years 1-10 = 0.15($26,300) = $3,945

$$0 = -\$110,000 + \$16,500 + (\$26,300 - \$3,945)(P/A,i\%,10) \qquad i = 20.1\%$$

12-50

With public utility bonds annual interest of $1,980 will be diminished by income taxes of 0.38($1,980) = $752.40. The resulting net positive cash flow from interest for years 1-8 = $1,980 - $752.40 = $1,227.60. Rate of return is $1,227.60 ÷ $15,000 = 0.0818 of 8.18%.

Income tax when municipal bonds mature will be 0.38(0.30)($20,000 - $15,000) = $570. Annual interest is 0.04($20,000) = $800.

$$0 = -\$15,000 + \$800(P/A,i\%,8) + (\$20,000 - \$570)(P/F,i\%,8) \qquad i = 8.11\%$$

(See Problem 14-35 for a further discussion of this comparison.)

12-51

David's combined incremental tax rate is 0.11 + 0.89(0.44) = 0.5016 or 50.16%. Therefore a $200 deductible gift will reduce his taxes by $100.32. The net cost of his gift will be $99.68 or approximately half the amount of the gift.

12-52

(a) Rate of return before income taxes:

0 = -\$20,000 + \$5,600(P/A,i%,10) i = 25.0%

(b) After-tax cash flows: (Headings same as Problem 12-1)

0	-\$20,000			+\$2,000	-\$18,000
1	+ 5,600	-\$5,600	\$ 0	0	+ 5,600
2-10	+ 5,600	- 1,600	+\$4,000	- 1,800	+ 3,800

0 = -\$18,000 + \$1,800(P/F,i%,1) + \$3,800(P/A,i%,10) i = 19.0%

(c)

0	-\$20,000			+\$2,000	-\$18,000
1-10	+ 5,600	-\$2,000	+\$3,600	- 1,620	+ 3,980

0 = -\$18,000 + \$3,980(P/A,i%,10) i = 17.8%

(d) Differences in cash flow are +\$1,620 at year 1 and -\$180 a year from years 2-10. With an i of 10% the present worth of this at date 0 is \$1,620(0.909) - \$180(6.144-0.909) = \$530

12-53

For the 20-year life equipment, the normal depreciation is \$20,000/20 = \$1,000. The accelerated depreciation is (\$20,000 - \$4,000)/20 = \$800 per year after depreciation of (\$4,000 + \$800) = \$4,800 in first year.

For the 10-year life equipment, the normal depreciation is \$20,000/10 = \$2,000/year. The accelerated depreciation is (\$20,000 - \$4,000)/10 = \$1,600 per year after the first-year depreciation of \$5,600.

A comparision of depreciation expenses shows:

	10-Year Life		20-Year Life			
	(1)	(2)	(3)	(4)		
Year	Normal	Accelerated	Normal	Accelerated	(1)+(4)	(2)+(3)
1	\$2000	\$5600	\$1000	\$4800	\$6800	\$6600
2-10	2000	1600	1000	800	2800	2600
11-20			1000	800	800	1000

It can be seen that using the accelerated depreciation for the 20-year life equipment allows greater depreciation deductions for tax purposes in the early years, and hence is superior to the other plan. The tax rate has no effect on the above solution and was thus omitted. The choice would be the same for any tax rate.

12-54

In the case of two used assets with the same lives and differing salvage values, one is indifferent to the selection of the asset to receive the accelerated depreciation. The total depreciation will be the same in either case.

	Zero Salvage Value		30% Salvage Value			
	(1)	(2)	(3)	(4)		
Year	Normal	Accelerated	Normal	Accelerated	(1)+(4)	(2)+(3)
1	$2000	$5600	$1400	$5000	$7000	$7000
2-10	2000	1600	1400	1000	3000	3000

Thus, the effect of depreciation on the taxable income will be the same regardless of which asset is chosen to receive the accelerated depreciation. The tax rate does not influence the amount of depreciation of the taxable income; therefore, it is irrelevant to the choice.

12-55

When income tax rates are expected to increase, the choice between more rapid and less rapid depreciation requires balancing the favorable effect of less taxes in the near future against unfavorable one of paying more taxes in the long run. As pointed out in the solution to Problem 12-41, this calls for a borrowing-financing type of decision based on recognition of the time value of money.

In this instance the tax rate on an increment of taxable income is expected to go up fairly rapidly, from 20% in year 1 to 46% in year 5. It would seem to be a reasonable guess that the most rapid methods of writeoff would not be attractive with the prospect of such a rapid increase in the applicable tax rate. Certainly it doesn't seem to be a good idea to use additional first year depreciation immediately while the applicable tax rate is only 20%.

The choice among different service lives and different depreciation methods really calls for a quantitative comparison rather than merely an intuitive analysis. Such a comparison must be based on specific estimates of applicable tax rates throughout a relevant study period. As explained in our solution to Problem 12-41, an appropriate criterion for choice is the cost of the money made available by choosing lives and depreciation methods that reduce the tax payments in the near future.

Problem 12-55 states that the corporation's applicable tax rate, which is now 20%, is expected to reach 46% within 5 years. A set of applicable tax rates consistent with this statement might be 20% for year 1, 30% for years 2 and 3, 40% for year 4, and 46% for year 5 and all subsequent years.

12-55 continued

If these tax rates are used in a comparison of depreciation methods and lives for the "certain assets" mentioned in the problem statement, analysis along the lines of our solution to Problem 12-41 suggests that straight-line depreciation with an 8-year life is superior both to years-digits depreciation with an 8-year life and to straight-line depreciation with a 12-year life. The cost of the money made available by the early tax savings from years-digits-8-year depreciation as compared to straight-line-8-year depreciation may be computed to be nearly 12% after taxes. In contrast, the cost of the money made available from choosing straight-line-8-year depreciation as compared to straight-line-12-year depreciation turns out to be slightly over 3% after taxes. No doubt an after-tax cost of money of 12% would seem too high in most cases; if so the straight-line method with an 8-year life would be preferable to the years-digits method with the same life. Also an after-tax cost of 3% presumably would be low enough to be attractive; if so the straight-line method with an 8-year life would be preferable to the same method with a 12-year life.

CHAPTER 13

Some Aspects of the Analysis of Multiple Alternatives

13-1

For an i* of 17%, Plan 7 is still the one to select. It is necessary to calculate the after-tax rate of return on its increment of investment over Plan 5; this turns out to be 17.4%. (Plan 6 is inacceptable as compared to Plan 5 because its incremental rate of return over Plan 5 is less than 17%.)

For an i* of 12%, Plan 7 continues to be the one to select. The incremental rate of return of Plan 8 over Plan 7 is only 10.0%; subsequent incremental rates are negative.

Of course no plan is acceptable with an after-tax i* of 20% because all of the rates of return in line H are less than 20%.

13-2

(All dollar figures that follow are in thousands, just as in Table 13-3.)

For Plan 7, depreciation used to determine taxable income will be \$1,512/20 = \$75.6 per year for years 1-20 and zero thereafter.

For years 1-20 annual income taxes will be 0.45(\$979.3 - \$333.1 - \$75.6) = \$256.8 and positive cash flow after income taxes will be \$389.4. For years 21-40 annual income taxes will be 0.45(\$979.3 - \$333.1) = \$290.8 and positive cash flow after income taxes will be \$355.4. Similar calculations for Plan 8 will show after-tax cash flow of \$441.7 for years 1-20 and \$398.0 for years 21-40. To find the rate of return on the extra investment of Plan 8 over Plan 7, solve:

$$0 = -(\$2{,}344 - \$1{,}912) + (\$441.7 - \$355.4)(P/A,i\%,20) + (\$398.0 - \$355.4)(P/A,i\%,40 - P/A,i\%,20)$$

i turns out to be 11.8%. Therefore the more favorable income tax treatment is not sufficient to justify the extra investment in Plan 8; Plan 7 still should be selected.

13-3

Table 13-4 shows CR costs of $403 and $484 for alternatives B-3 and B-4. With 100% terminal salvage values these figures become $350 and $420 (7% of the respective investments). The increment of benefits of B-4 over B-3 continues to be $70 as shown in Table 13-5. The increment of costs now is ($420 + $40 - $350 - $30) = $80. The incremental B/C ratio for B-4 over B-3 therefore is $70 ÷ $80 = 0.88 It follows that B-3 is preferable to B-4 even with the assumption of 100% terminal salvage value.

13-4

With an i* of 25% and an 8-year life the capital recovery factor is 0.3004.

Machine	Reduction in Annual Disbursements A	CR Cost B	A - B
Onondaga	$20,800	$16,220	$4,580
Oneida	22,700	18,380	4,320
Cayuga	26,700	21,630	5,070
Tuscarora	27,900	23,250	4,650
Seneca	32,100	27,580	4,520
Mohawk	37,300	32,440	4,860

With the 25% stipulation Cayuga should be selected. It maximizes (A - B). Its incremental before-tax rate of return over Onondaga is approximately 28.3%. The incremental before-tax rate of return of Mohawk over Cayuga is approximately 24.3%.

The following alternate solution is along the lines of Table 13-3:

	Onondaga	Oneida	Cayuga	Tuscarora	Seneca	Mohawk
A Investment	$54,000	$61,200	$72,000	$77,400	$91,800	$108,000
B Annual positive cash flow	20,800	22,700	26,700	27,900	32,100	37,300
C Rate of return on total investment	35.0%	33.5%	33.5%	32.3%	31.0%	30.5%
D Compare with		Onondaga	Onondaga	Cayuga	Cayuga	Cayuga
E Increment of investment		7,200	18,000	5,400	19,800	36,000
F Increment of annual cash flow		1,900	5,900	800	5,400	10,600
G Rate of return on increment of investment		20.5%	28.4%	4.1%	21.7%	24.3%
H Select		Onondaga	Cayuga	Cayuga	Cayuga	Cayuga

13-5

(A/P,15%,8) = 0.22285

	Onondaga	Oneida	Cayuga	Tuscarora	Seneca	Mohawk
A Annual cash flow before income taxes	$20,800	$22,700	$26,700	$27,900	$32,100	$37,300
B Annual depreciation for tax purposes	6,750	7,650	9,000	9,675	11,475	13,500
C Increase in taxable income (A - B)	14,050	15,050	17,700	18,225	20,625	23,800
D Increase in income tax (0.4C)	5,620	6,020	7,080	7,290	8,250	9,520
E Capital recovery @ 15%	12,034	13,638	16,045	17,249	20,458	24,068
F Excess of net annual positive cash flow over capital recovery @ 15%	3,146	3,052	3,575	3,361	3,392	3,712

With a 40% applicable tax rate and an i* of 15%, Mohawk is the best choice. Its incremental rate of return over Cayuga may be calculated to be approximately 15.5%.

13-6

Just as in Problems 13-4 and 13-5, the two leading competitors are Cayuga and Mohawk. The relevant rates of return for deciding regarding these two may be computed as follows:

In year 1 the years-digits depreciation for Cayuga is 8/36($72,000) = $16,000; the annual decrease is 1/36($72,000) = $2,000. The tax for year 1 will be 0.40($26,700-$16,000) = $4,280. The annual increase in tax will be 0.40($2,000) = $800. The comparable figures for Mohawk are, respectively, $24,000; $3,000; $5,320; and $1,200. Investment tax credits in year zero are 10% of the respective investments.

To find the after-tax rate of return for Cayuga, solve:

$$0 = -\$72,000 + 0.10(\$72,000) + (\$26,700-\$4,280)(P/A,i\%,8) - \$800(P/G,i\%,8)$$

A trial-and-error solution indicates that the rate of return is approximately 27.2%.

13-6 continued

To find the after-tax rate for Mohawk, solve:

$$0 = -\$108,000 + 0.10(\$108,000) + (\$37,300-\$5,320)(P/A,i\%,8) - \$1,200(P/G,i\%,8)$$

A trial-and-error solution indicates that the rate of return is approximately 24.9%.

To find the after-tax rate on the extra investment in Mohawk over Cayuga, solve:

$$o = -\$36,000 + 0.10(\$36,000) + (\$31,980-\$22,420)(P/A,i,8) - (\$1,200 - \$800)(P/G,i\%,8)$$

A trial-and-error solution indicates that the rate of return on the increment of investment is approximately 20.4%. Because the total return on Mohawk and this incremental rate of return both exceed the stipulated i* of 15%, Mohawk should be selected. It may be noted that this 20.4% is considerably better than the comparable incremental rate of return of 15.5% obtained in Problem 13-5 with less favorable tax treatment of the investment.

13-7

Annual costs:	n = 20	50	100
(A/P, 5%, n)	= 0.08024	0.05478	0.05038

Plan	EUAC
A:	\$5,000,000(0.05038) + \$60,000 = \$311,900
B:	\$311,900 + \$1,000,000(0.05478) + 25,000 = \$391,680
C:	\$311,900 + \$125,120 = \$437,020
D:	\$391,680 + \$125,120 = \$516,800
E:	\$3,750,000(0.05038) + \$50,000 = \$238,925
F:	\$238,925 + 125,120 = \$364,045
G:	(\$800,000-\$300,000)(0.08024)+\$300,000(0.05)+\$70,000 = \$125,12

Benefit/cost ratios:

	Total Benefits	Total EUAC	Ratio
A:	\$580,000	\$311,900	1.86
B:	760,000	391,680	1.94
C:	640,000	437,020	1.46
D:	820,000	516,800	1.59
E:	500,000	238,925	2.09
F:	590,000	364,045	1.62
G:	350,000	125,120	2.80

13-7 continued

Incremental B/C ratio analysis:

Defender	Challenger	Increment of Benefits	Increment of costs	Ratio	Winner
No project	G	$350,000	$125,120	2.80	G
G	E	150,000	113,805	1.32	E
E	A	80,000	72,975	1.10	A
A	F	10,000	52,145	0.19	A
A	C	60,000	125,120	0.48	A
A	B	180,000	79,780	2.26	B
B	D	60,000	125,120	0.48	B

Select Plan B, Willow dam, reservoir, and power plant.

13-8

Annual costs	n = 20	n = 50	n = 100
(A/P,8%,n)	0.10185	0.08174	0.08004

Plan	EUAC
A	$5,000,000(0.8004) + $60,000 = $460,200
B	$460,200 + $1,000,000(0.08174) + $25,000 = $566,940
C	$460,200 + $144,925 = $605,125
D	$566,940 + $144,925 = $711,865
E	$3,750,000(0.08004) ÷ $50,000 = $350,150
F	$350,150 + $144,925 = $495,075
G	($800,000 - $300,000)(0.10185) + $300,000(0.08) + $70,800 = $144,925

Benefit/cost ratios:

	Total Benefits	Total EUAC	Ratio
A	$580,000	$460,200	1.26
B	760,000	566,940	1.34
C	640,000	605,125	1.06
D	820,000	711,865	1.15
E	500,000	350,150	1.43
F	590,000	495,075	1.19
G	350,000	144,925	2.42

13-8 continued

Incremental B/C ratio analysis:

Defender	Challenger	Increment of Benefits	Increment of Costs	Ratio	Winner
No project	G	$350,000	$144,925	2.42	G
G	E	150,000	205,225	0.73	G
G	A	230,000	315,275	0.73	G
G	F	240,000	350,150	0.69	G
G	C	290,000	460,200	0.63	G
G	B	410,000	422,015	0.97	G
G	D	470,000	566,940	0.83	G

Select G, channel improvement alone.

13-9

$$a = \$1.90(0.00302)(0.20182) = \$0.001158$$

$$b = \frac{(40)^2(3{,}500)(\$0.072)(10{,}580)}{1{,}000} = \$4{,}265{,}856$$

$$x_e = \frac{b}{a} = \frac{\$4{,}265{,}856}{\$0.001158} = 60{,}700 \text{ circular mils}$$

This falls between size 2 (66,400 circular mils) and size 3 (52,600 circular mils)

13-10

A Size of wire (AWG)	00	0	1	2	3
B Weight of wire in lb.	403	319	253	201	159
C Investment in wire	$765.70	$606.10	$480.70	$381.90	$302.10
D Resistance in ohms	0.0795	0.100	0.126	0.159	0.201
E Power loss - kw	0.1272	0.1600	0.2016	0.2544	0.3216
F Annual energy loss in kw-hr	445	560	706	890	1,126
G Annual investment charges @ 20.182%	$154.53	$122.32	$ 97.01	$ 77.08	$ 60.97
H Annual cost of lost energy @ 7.2¢/kw-hr	$ 32.04	$ 40.32	$ 50.83	$ 64.08	$ 81.07
I Equivalent uniform annual cost assumed to be variable with wire size	$186.57	$162.64	$147.84	$141.16	$142.04

Size 2 has the lowest annual cost

13-11

A Size of wire	00	0	1	2	3
B CR @ 7% (0.08581)	$60.52	$47.90	$37.99	$30.18	$23.88
C Property tax (0.0175)	12.34	9.77	7.75	6.16	4.87
D Depreciation	28.21	22.33	17.71	14.07	11.13
E Cost of lost energy	28.39	35.70	44.98	56.78	71.77
F Deductions from taxable income (C + D + E)	68.94	67.80	70.44	77.01	87.77
G Extra annual income taxes above size 3	9.42	9.98	8.66	5.38	0.00
H Equivalent uniform annual cost assumed to be variable with wire size	110.67	103.35	99.38	98.50	100.52

Size 2 still shows the lowest annual cost. However, the differences in annual costs among the different wire sizes are considerably less than in the before-tax analysis of Table 13-2.

13-12

Because the 10% tax credit, in effect, reduces the investment by 10%, it reduces the annual capital recovery costs (line B in the foregoing table) by 10%. If total annual costs (line H) are reduced by 10% of line B, the resulting figures are:

00	$104.62
0	98.56
1	95.58
2	95.48
3	98.13

The investment tax credit makes comparative annual costs almost the same for sizes 1 and 2.

13-13 .

A Size of Wire	00	0	1	2	3
B Weight, lbs.	403	319	253	201	159
C Investment @$2.25/lb	$906.75	$717.75	$569.25	$452.25	$357.75
D Resistance, ohms	0.0795	0.100	0.126	0.159	0.201
E Power loss, 60_A, kw	0.2862	0.3600	0.4536	0.5724	0.7236
F Annual energy loss, kw. hr.	1,717	2,160	2,722	3,434	4,342
G. Investment charges @ 18.5%	$167.75	$132.78	$105.31	$ 83.67	$ 66.18
H Cost of lost energy	72.12	90.72	114.31	144.24	182.35
I Total annual cost assumed to be variable with the size of wire	$239.87	$223.51	$219.62	$227.91	$248.53

Size 1 is most economical, but a very small increase will provide for greater growth in demand.

13-14

Proposal	Required Investment	Annual Savings in Dir. Materials and Labor	Gross Return	Incremental Return over next lower acceptable alternativ
A	$6,000	$1,500	25.0%	----Eliminate A
B	8,000	2,500	31.3	
C	12,000	4,100	34.2	40%>30%, keep C
D	13,000	4,200	32.3	10%<30%, eliminate D
E	18,000	6,000	33.3	31.7%(for E over C) keep E
F	25,000	7,600	30.4	22.9%<30%, eliminate

Proposal E is the largest investment that will meet the requirement that it earn a gross return of 30% as well as that each increment of investment earns a gross return of 30% or more.

13-15

CR + Annual Maintenance Costs:

Site A: $6,000,000(A/P,5.5%,75) + $100,000 = $436,060
Site B: $8,000,000(A/P,5.5%,75) + $140,000 = $588,080
CI : $1,000,000(A/P,5.5%,25) + $230,000 = $304,550
A + CI: $436,060 + $304,550 = $740,610
B + CI: $588,080 + $304,550 = $892,630

13-15 continued

Arrange the five plans in increasing order of investment required:

Plan	Annual Benefit	Annual Cost	B/C Ratio	Increment	Increment of Benefit	Increment of cost	B/C
CI	$680,000	$304,550	2.23	--	$680,000	$304,550	
A	820,000	436,060	1.88	A/CI	140,000	131,510	1.06
B	940,000	588,080	1.60	B/A	120,000	152,020	0.79
		Eliminate B					
A+CI	1,000,000	740,610	1.35	A+CI/A	180,000	304,550	0.59
		Eliminate A+CI					
B+CI	1,080,000	892,630	1.21	B+CI/A	260,000	456,570	0.57
		Eliminate B+CI					

Choose Site A alone, because it is the largest investment that meets the criterion of a B/C ratio equal to or greater than 1 for each increment of investment.

Computation of the total equivalent uniform annual cost, including remaining flood damage, will lead to the same conclusion.

13-16

CR + annual maintenance costs:

Site A: $6,000,000(A/P,8%,75) + $100,000 = $581,500
Site B: $8,000,000(A/P,8%,75) + $140,000 = $782,000
CI : $1,000,000(A/P,8%,25) + $230,000 = $323,680
A + CI: $581,500 + $323,680 = $905,180
B + CI: $782,000 + $323,680 = $1,105,680

Arrange the five plans in increasing order of investment required.

Plan	Annual Benefits	Annual Cost	B/C Ratio	Increment	Incr. of Benefits	Incr. of Costs	B/C
CI	$1,360,000	$323,680	4.20				
A	1,640,000	581,500	2.82	A/CI	$280,000	$257,820	1.09
B	1,880,000	782,000	2.40	B/A	240,000	200,500	1.20
A+CI	2,000,000	905,180	2.21	A + CI/B	120,000	123,180	0.97
		Eliminate A + CI					
B+CI	2,160,000	1,105,680	1.95	B + CI/B	280,000	323,680	0.87
		Eliminate B + CI					

Choose Site B alone.

13-17

Examination of the gross returns on the incremental investments leads one to eliminate proposals A and E immediately. Eliminating A from consideration will not affect the gross returns shown for B, C, and D. For each of these proposals, the gross return on the extra investment over the next lower proposal (in terms of investment) is unchanged. However, the elimination of E from consideration requires a new computation for F. The extra investment of F over D = $22,000. The incremental annual savings of F over D = $4,900. The gross return on the extra investment in F over D = 22.3%. This does not meet the criterion, so F is eliminated from consideration. D should be chosen as the largest investment for which the gross return on each avoidable increment of investment was equal to or greater than 25%, the criterion. Smith, Jones, and Johnson all used the wrong interpretation of the data for decision making.

13-18

Net investments after investment tax credit are:

Cast iron	$3,500
Alloy A	$6,500 - (6-2/3% of $6,500) = $6,067
Alloy B	$11,000 - (10% of $11,000) = $9,900

Extra annual income tax for Alloy A over cast iron

$$= 0.45\left(\frac{\$3,500}{2} - \frac{\$6,500}{5}\right) = \$202$$

Extra annual income tax for Alloy B over cast iron

$$= 0.45\left(\frac{\$3,500}{2} - \frac{\$11,000}{10}\right) = \$292$$

Comparing the sum of capital recovery costs at i* = 10% plus extra income tax:

Cast iron:	$3,500(A/P,10%,2) = $2,017
Alloy A:	$6,067(A/P,10%,5) = $1,802
Alloy B:	$9,900(A/P,10%,10) = $1,903

Alloy A has the lowest comparative annual cost with an after-tax i* of 10%.

13-19

Plan A

0 = \$176,000 - \$55,000 - \$200,000i - \$400,000(A/P,i%,40)
i = 20.2% Reject Plan A because prospective rate of return before income taxes is less than the 25% standard.

Plan B

0 = \$260,000 - \$70,000 - \$200,000i - \$560,000(A/P,i%,40)
i = 25.0% Plan B is acceptable but must defend itself against the proposals that call for larger investments.

Plan C

0 = \$380,000 - \$100,000 - \$200,000i - \$800,000(A/P,i%,40)
i = 28.0%

To compare Plan C with Plan B

0 = \$90,000 - \$240,000(A/P,i%,40)
i = 37.5% Plan C is preferable to Plan B. Its return on total investment and its return on the incremental investment both exceed the 25% standard.

Plan D

0 = \$467,000 - \$125,000 - \$200,000i - \$1,080,000(A/P,i%,40)
i = 26.7%

To compare Plan D with Plan C

0 = \$62,000 - \$280,000(A/P,i%,40)
i = 22.1% Plan D should be rejected because its incremental rate of return over Plan C is less than the 25% standard.

Plan E

0 = \$598,000 - \$150,000 - \$200,000i - \$1,520,000(A/P,i%,40)
i = 26.0%

To compare Plan E with Plan C

0 = \$168,000 - \$720,000(A/P,i%,40)
i = 23.3% Plan E should be rejected because its incremental rate of return over Plan C is less than the 25% standard.

Select Plan C.

13-20

	Annual taxable income	Annual tax @ 40%
Plan A	\$176,000 - \$55,000 - \$10,000 = \$111,000	\$ 44,000
Plan B	\$260,000 - \$70,000 - \$14,000 = \$176,000	\$ 70,400
Plan C	\$380,000 - \$100,000 - \$20,000 = 260,000	\$104,000
Plan D	\$467,000 - \$125,000 - \$27,000 = 315,000	\$126,000
Plan E	\$598,000 - \$150,000 - \$38,000 = 410,000	\$164,000

13-20 continued

Plan A

0 = $121,000 - $44,000 - $200,000i - $480,000(A/P,i%,40)
i = 12.8% Reject Plan A because prospective rate of return after income taxes is less than the 15% standard.

Plan B

0 = $190,000 - $70,400 - $200,000i - $560,000(A/P,i%,40)
i = 15.7% Plan B is acceptable but must defend itself against proposals that call for larger investments.

Plan C

0 = $280,000 - $104,000 - $200,000i - $800,000(A/P,i%,40)
i = 17.6%

To compare Plan C with Plan B

0 = $56,400 - $240,000(A/P,i%,40)
i = 23.5% Plan C is preferable to Plan B. Its return on total investment and its return on the incremental investment both exceed the 15% standard.

Plan D

0 = $342,000 - $126,000 - $200,000i - $1,080,000(A/P,i%,40)
i = 16.8%

To compare Plan D with Plan C

0 = $40,000 - $280,000(A/P,i%,40)
i = 14.2% Plan D should be rejected because its incremental rate of return over Plan C is less than the 15% standard.

Plan E

0 = $448,000 - $164,000 - $200,000i - $1,520,000(A/P,i%,40)
i = 16.5%

To compare Plan E with Plan C

0 = $108,000 - $720,000(A/P,i%,40)
i = 14.94% This incremental rate of return is very close to the stipulated i* of 15% after income taxes. The choice between Plans C and E doubtless would call for a critical look at the irreducibles, which were not mentioned in the problem statement.

13-21

	(a)	(b)	(c)	(d)	(e)	(f)	(g)
Plan A	\$121,000	\$26,667	\$94,333	\$37,733	\$48,400	\$83,267	\$72,600
Plan B	190,000	37,333	152,667	61,067	76,000	128,933	114,000
Plan C	280,000	53,333	226,667	90,067	112,000	189,933	168,000
Plan D	342,000	72,000	270,000	108,000	136,800	234,000	205,200
Plan E	448,000	101,333	346,667	138,667	179,200	309,333	268,800

(a) Annual before-tax cash flow. This will be taxable income in years 16-40 when buildings are fully depreciated and is taxable income before allowable depreciation in years 1-15.
(b) Depreciation allowable, years 1-15
(c) Taxable income, years 1-15 -- (a) - (b)
(d) Income tax at 40%, years 1-15 -- 0.40(c)
(e) Income tax at 40%, years 16-40 -- 0.40(a)
(f) After-tax cash flow, years 1-15 -- (a) - (d)
(g) After-tax cash flow, years 16-40 -- (a) - (e)

Plan A

$$0 = -\$600,000 + \$72,600(P/A,i\%,40) + (\$83,267 - \$72,600)(P/A,i\%,15) + \$200,000(P/F,i\%,40)$$

$i = 13.6\%$ Reject Plan A because prospective rate of return after income taxes is less than 15% standard.

Plan B

$$0 = -\$760,000 + \$114,000(P/A,i\%,40) + \$14,933(P/A,i\%,15) + \$200,000(P/F,i\%,40)$$

$i = 16.8\%$ Plan B is acceptable but must defend itself against proposals that call for larger investments.

Plan C

$$0 = -\$1,000,000 + \$168,000(P/A,i\%,40) + \$21,933(P/A,i\%,15) + \$200,000(P/F,i\%,40)$$

$i = 18.9\%$

To compare Plan C with Plan B

$$0 = -\$240,000 + \$54,000(P/A,i\%,40) + \$7,000(P/A,i\%,15)$$

$i = 25.4\%$ Plan C is preferable to Plan B. Its return on total investment and its return on the increment of investment both exceed the 15% standard.

Plan D

$$0 = -\$1,280,000 + \$205,200(P/A,i\%,40) + \$28,800(P/A,i\%,15) + \$200,000(P/F,i\%,40)$$

$i = 18.1\%$

13-21 continued

To compare Plan D with Plan C

$$0 = -\$280,000 + \$37,200(P/A,i\%,40) + \$6,867(P/A,i\%,15)$$

i = 15.4% Plan D is preferable to Plan C. Its return on total investment and its return on the increment of investment both exceed the 15% standard.

Plan E

$$0 = -\$1,720,000 + \$268,800(P/A,i\%,40) + \$40,533(P/A,i\%,15) + \$200,000(P/F,i\%,40)$$

i = 17.8% Plan E is preferable to Plan D. Its return on total investment and its return on the increment of investment both exceed the 15% standard.

Select Plan E.

13-22

Proposal	Annual highway-user costs	CR of 50-year investment	CR of 10-year investment	Annual highway upkeep	Total annual cost
R-0	$920,000	$ 0	$ 0	$100,000	$1,020,000
R-1	800,000	0	104,320	70,000	974,320
R-2	760,000	40,870	104,320	70,000	975,190
R-3	700,000	114,440	89,420	60,000	963,860
T-1	630,000	245,220	81,970	55,000	1,012,190
T-2	580,000	326,960	74,520	50,000	1,031,480

Proposal R-3 has the lowest annual cost with an i* of 8%.

13-23

(same headings as in Problem 13-22)

R-0	$920,000	$ 0	$ 0	$100,000	$1,020,000
R-1	800,000	0	82,060	70,000	952,060
R-2	760,000	19,440	82,060	70,000	931,500
R-3	700,000	54,420	70,340	60,000	884,760
T-1	630,000	116,610	64,480	55,000	866,090
T-2	580,000	155,480	58,620	50,000	844,100

Proposal T-2 has the lowest annual cost with an i* of 3%.

13-24

(same headings as in Problem 13-22)

R-0	$920,000	$ 0	$ 0	$100,000	$1,020,000
R-1	800,000	0	139,480	70,000	1,009,480
R-2	760,000	75,070	139,480	70,000	1,044,550
R-3	700,000	210,200	119,550	60,000	1,089,750
T-1	630,000	450,420	109,590	55,000	1,245,010
T-2	580,000	600,560	99,620	50,000	1,330,180

Proposal R-1 has the lowest annual cost with an i* of 15%.

13-25

Cash flow comparison, one 2-year vs. two 1-year subscriptions

Year	2-year	two 1-year	Difference
0	-$40	-$25	-$15
1		- 25	+ 25

Rate of return is $\frac{\$25 - \$15}{\$15} = 0.67$ or 67% on the extra investment required for the 2-year subscription.

Compare cash flow for one 3-year with three 1-year subscriptions

0	-$50	-$25	-$25
1		- 25	+ 25
2		- 25	+ 25

Calculation from the formula shows the rate of return is nearly 62%.

CHAPTER 14

Prospective Inflation and

Sensitivity Analysis

14-1

$0 = -\$9{,}600 + \$400(P/A,i\%,10) + \$10{,}000(P/F,i\%,10)$

i = 4.5% before the capital gains tax with no consideration of inflation.

The following analysis along the lines of Table 14-2 (p.317) includes the effect of inflation and of the $100 capital gains tax in 1970.

Year	Cash flow current dollars	Consumer Price Index	Cash flow converted to 1960 dollars
1960	-$9,600	88.7	-$9,600
1961	+400	89.6	+396
1962	+400	90.6	+392
1963	+400	91.7	+387
1964	+400	92.9	+382
1965	+400	94.5	+375
1966	+400	97.2	+365
1967	+400	100.0	+355
1968	+400	104.2	+340
1969	+400	109.8	+323
1970	+10,300	116.3	+7,856

The present worth of the series in the right-hand column is zero with an i of approximately 1.8%.

14-2

With 8% inflation the after-tax cash flow in zero-date dollars will be:

0	-$10,000	4	+$617	8	+$454
1	+778	5	+572	9	+420
2	+720	6	+529	10	+5,021
3	+667	7	+490		+$268

The present worth of the foregoing series is zero with an i of approximately 0.4%.

14-3

With 4% inflation the after-tax cash flow in zero-date dollars will be:

0	-$10,000	4	+$718	8	+$614
1	+808	5	+690	9	+590
2	+777	6	+664	10	+7,324
3	+747	7	+638		+$3,570

The present worth of the foregoing series is zero with an i of approximately 4.2%.

14-4

Year	Cash flow before income taxes in zero-date dollars	Estimated price index assuming 100 for zero-date	Estimated before-tax cash flow in current dollars	Write-off of initial outlay for tax purposes	Taxable income from project	Cash flow for income taxes in current dollars	Cash flow after income taxes in current dollars	Cash flow after income taxes in zero-date dollars
			BA/100		(C + D)	-0.49E	(C + F)	100G/B
	A	B	C	D	E	F	G	H
0	-$110,000	100.00	-$110,000				-$110,000	-$110,000
1	+26,300	112.00	+29,456	-$11,000	+$18,456	-$9,043	+20,413	+18,226
2	+26,300	125.44	+32,991	-11,000	+21,991	-10,776	+22,215	+17,710
3	+26,300	140.49	+36,949	-11,000	+25,949	-12,715	+24,234	+17,250
4	+26,300	157.35	+41,383	-11,000	+30,383	-14,888	+26,495	+16,838
5	+26,300	176.23	+46,348	-11,000	+35,348	-17,321	+29,027	+16,471
6	+26,300	197.38	+51,911	-11,000	+40,911	-20,046	+31,865	+16,144
7	+26,300	221.07	+58,141	-11,000	+47,141	-23,049	+35,042	+15,851
8	+26,300	247.60	+65,119	-11,000	+54,119	-26,518	+38,601	+15,590
9	+26,300	277.31	+72,933	-11,000	+61,933	-30,347	+42,586	+15,357
10	+26,300	310.58	+81,683	-11,000	+70,683	-34,635	+47,048	+15,148

The present worth of the series in the right hand column is zero with an i of approximately 8.4%.

14-5

Headings correspond to those in Problem 14-4.

Year	A	B	C	D	E	F	G	H
0	-$110,000	100.00	-$110,000				-$110,000	-$110,000
1	+26,300	106.00	+27,878	-$4,400	+$23,478	-$11,504	+16,374	+15,447
2	+26,300	112.36	+29,551	-4,400	+25,151	-12,324	+17,227	+15,332
3	+26,300	119.10	+31,323	-4,400	+26,923	-13,192	+18,131	+15,223
4	+26,300	126.25	+33,204	-4,400	+28,804	-14,114	+19,090	+15,121
5	+26,300	133.82	+35,195	-4,400	+30,795	-15,090	+20,105	+15,024
6	+26,300	141.35	+37,307	-4,400	+32,907	-16,124	+21,183	+14,933
7	+26,300	150.36	+39,545	-4,400	+35,145	-17,221	+22,324	+14,847
8	+26,300	159.38	+41,917	-4,400	+37,517	-18,383	+23,534	+14,766
9	+26,300	168.95	+44,434	-4,400	+40,034	-19,617	+24,817	+14,689
10	+26,300	179.08	+47,098	-4,400	+42,698	-20,922	+26,176	+14,617
11		189.83		-4,400	-4,400	+2,156	+2,156	+1,136
12		201.22		-4,400	-4,400	+2,156	+2,156	+1,071
13		213.29		-4,400	-4,400	+2,156	+2,156	+1,011
14		226.09		-4,400	-4,400	+2,156	+2,156	+954
15		239.66		-4,400	-4,400	+2,156	+2,156	+900
16		254.04		-4,400	-4,400	+2,156	+2,156	+849
17		269.28		-4,400	-4,400	+2,156	+2,156	+801
18		285.43		-4,400	-4,400	+2,156	+2,156	+755
19		302.56		-4,400	-4,400	+2,156	+2,156	+713
20		320.71		-4,400	-4,400	+2,156	+2,156	+672
21		339.96		-4,400	-4,400	+2,156	+2,156	+634
22		360.35		-4,400	-4,400	+2,156	+2,156	+598
23		381.97		-4,400	-4,400	+2,156	+2,156	+564
24		404.89		-4,400	-4,400	+2,156	+2,156	+532
25		429.19		-4,400	-4,400	+2,156	+2,156	+502

The present worth of the series in the right-hand column is zero with an i of approximately 6.9%.

14-6

Headings correspond to those in Problem 14-4.

Year	A	B	C	D	E	F	G	H
0	-$110,000	100.00	-$110,000				-$110,000	-$110,000
1	+26,300	112.00	+29,456	-$22,000	+$7,456	-$3,653	+25,803	+23,038
2	+26,300	125.44	+32,991	-22,000	+10,991	-5,386	+27,605	+22,007
3	+26,300	140.49	+36,949	-22,000	+14,949	-7,325	+29,624	+21,086
4	+26,300	157.35	+41,383	-22,000	+19,383	-9,498	+31,885	+20,262
5	+26,300	176.23	+46,348	-22,000	+24,348	-11,931	+34,417	+19,530
6	+26,300	197.38	+51,911	0	+51,911	-25,436	+26,475	+13,413
7	+26,300	221.07	+58,144	0	+58,144	-28,489	+29,652	+13,413
8	+26,300	247.60	+65,119	0	+65,119	-31,908	+33,211	+13,413
9	+26,300	277.31	+72,933	0	+72,933	-35,737	+37,196	+13,413
10	+26,300	310.58	+81,683	0	+81,683	-40,025	+41,658	+13,413

The present worth of the series in the right-hand column is zero with an i of approximately 10.6%.

14-7

In Example 9-1 annual road user costs were \$26,700 lower with location K than with J; other annual costs were \$46,500 higher with K. If the difference in road used costs should increase immediately by 75% the effect would be to make K break even with J as long as there were no changes in other costs. If changes that would increase this difference by more than 75% are anticipated, the relevant question is how soon they are expected to take place.

14-8

Of course the prospect of an upward differential change in the value of energy favors the two plans involving a power plant, namely, Plans B and D. Because B and D will produce equal amounts of hydroelectric power, the prospect of the differential change is neutral as between B and D. The solution of Problem 13-7 favored the selection of Plan B; the prospect of an upward differential change in the value of energy obviously strengthens the case for this plan.

(This problem refers to Problem 13-7, in which i* was 5%, and does not mention 13-8, in which i* was 8%. It may be noted that in 13-8 the incremental B/C ratio for Plan B over Plan G was 0.97; not much prospective differential change in the value of electric energy would be needed to raise this ratio above 1.0 and to change the economic choice from G to B.)

14-9

Without the prospect of differential price change the prospect of inflation ought not to have been an important matter. But the prospect of upward differential price changes in, say, the price of power plant fuel and the price of purchased electricity might reasonably have influenced the choice; it seems likely that these two changes would not have equal influence on the alternatives. Purchased electricity presumably will be a larger proportion of total disbursements than purchased fuel. Conceivably the prospect of these two differential changes might be enough to change the economic choice from purchase to generation.

14-10

(a) $0 = -\$100{,}000 + \$300{,}000(P/F,i\%,10)$

i = approximately 11.6%, the apparent rate of return before income taxes if the effect of inflation is not recognized.

(b) On date 10 the tax on the so-called profit is 0.45(\$300,000 - \$100,000) = \$90,000. The net cash flow after this tax is \$300,000 - \$90,000 = \$210,000

$0 = -\$100{,}000 + \$210{,}000(P/F,i\%,10)$

i = approximately 7.7%, the apparent rate of return after income taxes if the effect of inflation is neglected.

(c) Because the price level has doubled, the cash flow at date 10 must be multiplied by 0.5 to convert it to zero-date dollars.

$0 = -\$100{,}000 + \$210{,}000(0.5)(P/F,i\%,10)$

i = approximately 0.5% after taxes if the effect of inflation is recognized.

14-11

Of course the answer to (a) is unchanged.

(b) The tax now is 0.15(\$300,000 - \$100,000) = \$30,000

$0 = -\$100{,}000 + \$270{,}000(P/F,i\%,10)$

i = approximately 10.5%

(c) $0 = -\$100{,}000 + \$270{,}000(0.5)(P/F,i\%,10)$

i = approximately 3.0%

14-12

In 14-10(c) the tax will be 0.45(\$100,000) = \$45,000

$0 = -\$100{,}000 + \$255{,}000(0.5)(P/F,i\%,10)$

i = approximately 2.5%

In 14-11(c) the tax will be 0.15(\$100,000) = \$15,000

$0 = -\$100{,}000 + \$285{,}000(0.5)(P/F,i\%,10)$

i = approximately 3.6%

14-13

PW of residual, R, for i = 7%, n = 20, = R(R/F,7%,20) = 0.2584R

(a) 50% increase: Compare J with H:
B/C = \$706,000/[\$364,000 - 0.2584(0.5)(\$330,000)]= 2.2

Compare K with J:

B/C = \$282,000/[\$493,000 - 0.2584(0.5)(\$550,000 - \$330,000)]
= 0.61

(b) 50% decrease: Compare J with H:
B/C = \$706,000/[\$364,000 + 0.2584(0.5)(\$330,000)] = 1.75
Compare K with J:
B/C = \$282,000/[\$493,000 + 0.2584(0.5)(\$550,000 - \$330,000)]
= 0.54

14-14

Let x represent the first cost of a tank with Material B that will have the same annual cost as a tank with Material A. Equating the annual costs:

x(A/P,25%,15) = \$35,000(A/P,25%,6) + \$12,000(P/F,25%,3)·(A/P,25%,6)

x = \$53,800

14-15

Solution omitted.

14-16

PW of costs for 15 years	2,500 Gal/Day	5,000 Gal/Day
First cost	#32,000	#48,000
Replacement		
#32,000(P/F,15%,5) =	15,846	
Less PW of residual value		
#32,000(A/P,15%,15)(P/A,15%,5)(P/F,15%,15) =	-2,255	
Operation, Maintenance, & Repairs:		
#1,000(P/A,15%,15) - #500(P/A,15%,5) =	4,171	
#750(P/A,15%,15) =		4,385
Electric Energy:		
#20[1,000(P/A,15%,15) + #350(P/G,15%,11) +#3,500(P/A,15%,4)(P/F,15%,11)]365/100,000 =	1,072	
#15(14,690)365/100,000 =		804
Chlorine:		
#350(365)(14,690)/150,000 =	12,511	12,511
Diatomacious Earth:		
#60(365)(14,690)/75,000 =	4,289	4,289
Total	#67,634	#69,989

Sensitivity to interest rate:		
PW at 20%	#59,925	#64,538
PW at 15% (from above)	67,634	69,989
PW at 10%	78,986	78,552

As we can see from these figures, increasing the i* will not change the indicated decision, but a reduction to 10% could change it. Sensitivity to rate of growth of demand:

At an increase of 450 gal/day per year	69,962	72,282
" " " " 350 gal/day per year	67,634	69,989
" " " " 250 gal/day per year	65,308	67,698

The result is the conclusion that variation in the rate of growth in demand does not tend to reverse the decision.

14-17

Year	Taxable income	Tax	Tax percentage	Tax rate on increment of investment
0	$45,000	$8,250	18.33%	20%
1	51,750	9,775	18.89%	30%
2	59,512	12,104	20.34%	30%
3	68,439	14,782	21.60%	30%
4	78,705	18,232	23.16%	40%
5	90,511	22,954	25.36%	40%
6	104,088	28,630	27.51%	46%

14-18

Under the stated conditions a 50% rise in the price level would lead to a taxable income of $60,000 with an incremental tax rate of 30%. A rise of 100% would lead to a taxable income of $80,000 with an incremental tax rate of 40%. A rise of 150% would lead to $100,000. Although the final increment of income would be taxed at 40%, any additional income would be taxed at 46%.

14-19

(a) Each figure in Table 12-1 will be multiplied by $(1.12)^n$ to translate it into current-year dollars and then divided by $(1.12)^n$ to translate it back to zero-date dollars. The zero-year dollars thus derived obviously will be identical with those in Table 12-1 and the before-tax rate of return will continue to be 20.0%.

(b) The tax on the so-called capital gain will be (0.49)($341,600 - $110,000) = $118,384. The after-tax cash flow from the sale of the land at the end of year 10 will be $341,600 - $118,384 = $223,216. Expressed in zero-date dollars this is

$\$223,216/(1.12)^{10} = \$71,876.$

$0 = -\$110,000 + \$11,220(P/A,i\%,10) + \$71,876(P/F,i\%,10)$

i = approximately 7.8%, the prospective after-tax rate of return with all cash flow expressed in dollars that have zero-date purchasing power.

(c) The capital-gains tax is now (0.20)($341,600 - $110,000) = $48,320. The after-tax cash flow from the sale of the land after 10 years becomes $341,600 - $48,320 = $293,280. Expressed in zero-date dollars this is $94,436.

$0 = -\$110,000 + \$11,220(P/A,i\%,10) + \$94,436(P/F,i\%,10)$
i = approximately 9.3%

14-20

(a) $\$550{,}000/(1.12)^{10} = \$177{,}100$

$0 = -\$110{,}000 + \$22{,}000(P/A,i\%,10) + \$177{,}100(P/F,i\%,10)$

i = approximately 22.3%

(b) Capital-gains tax will be 0.49($550,000 - $110,000) = $215,600. The after-tax cash flow from the sale of the land at the end of year 10 becomes $550,000 - $215,600 = $334,400. Expressed in zero-date dollars, this is $107,677.

$0 = -\$110{,}000 + \$11{,}220(P/A,i\%,10) + \$107{,}677(P/F,i\%,10)$

i = approximately 10.1%

(c) Capital-gains tax will be 0.20($550,000 - $110,000) = $88,000. The after-tax cash flow from the sale of the land at the end of year 10 becomes $550,000 - $88,000 = $462,000. Expressed in zero-date dollars, this is $148,764.

$0 = -\$110{,}000 + \$11{,}220(P/A,i\%,10) + \$148{,}764(P/F,i\%,10)$

i = approximately 12.2%

14-21

(a) $\$220{,}000/(1.12)^{10} = \$70{,}840$

$0 = -\$110{,}000 + \$22{,}000(P/A,i\%,10) + \$70{,}840(P/F,i\%,10)$

i = approximately 18.6%

(b) Capital-gains tax will be 0.49($220,000 - $110,000) = $53,900. The after-tax cash flow from the sale of the land at the end of year 10 becomes $220,000 - $53,900 = $166,100. Expressed in zero-date dollars, this is $53,484.

$0 = -\$110{,}000 + \$11{,}220(P/A,i\%,10) + \$53{,}484(P/F,i\%,10)$

i = approximately 6.4%

(c) Capital-gains tax will be 0.20($220,000 - $110,000) = $22,000. The after-tax cash flow from the sale of the land at the end of year 10 becomes $220,000 - $22,000 = $198,000. Expressed in zero-date dollars, this is $63,756.

$0 = -\$110{,}000 + \$11{,}220(P/A,i\%,10) + \$63{,}756(P/F,i\%,10)$

i = approximately 7.2%

14-22

The following matrix shows the relevant figures for calculated rates of return

	Conventional analysis -no inflation considered	Analysis in terms of zero-date dollars with 12% annual inflation and differential price change only in resale price of land		
		No d.p.c.	Favorable d.p.c.	Unfavorable d.p.c.
Before taxes	20.0%	20.0%	22.3%	18.6%
After taxes with 49% tax rate on all taxable income	10.2%	7.8%	10.1%	6.4%
After taxes with 49% tax rate on ordinary income and 20% rate on capital gains	10.2%	9.8%	12.2%	7.2%

To begin with, it should be noted that Problems 14-19, 14-20, and 14-21 do not involve the distortions introduced into the calculation of taxable income that are caused by basing depreciation deductions on costs at a different price level. Because Example 12-1 did not involve any assets depreciable for tax purposes, depreciation does not enter into these three problems based on this example. Moreover, because it was assumed that all other relevant cash flows would increase at the same rate as general price levels, the only differential price changes are those from the sale of the land at the end of the 10 years and from the income taxes based on that sale.

Therefore, these three problems concentrate attention on what happens when taxable capital gains are calculated without recognition of changes in the value of the monetary unit. The adverse effect of this on investors during a period of substantial inflation is apparent. Also it is evident that there is an equitable basis for taxing capital gains computed in this usual way at a lower rate (20% in these problems) than the rate applied to ordinary income (49% in these problems).

14-23

(a) The $1,000 a year positive gradient in cash flow due to the reduction in labor costs will be partially offset by a negative gradient of $490 a year caused by increased income taxes.
The equation to find before-tax rate of return becomes:

$$0 = -\$110,000 + \$26,300(P/A,i\%,10) + \$1,000(P/G,i\%,10)$$

i = approximately 23.8% before income taxes.

The equation to find after-tax rate of return becomes:

$$0 = -\$110,000 + \$23,213(P/A,i\%,10) + (\$1,000-\$490-\$980)(P/G,i\%,10)$$

i = approximately 14.6%

The prospect of the favorable differential price change increases the prospective before-tax rate of return by 3.7% (i.e., 23.8% - 20.1%) and increases the prospective after-tax rate by 2.3% (i.e., 14.6% - 12.3%).

(b) The answer to this question depends on whether the prosective before-tax gradient of $1,000 a year is an expected differential price change or whether it is expected to take place in part or entirely because of general inflation. The answer in (a) would not be changed if a differential price change is expected. If the gradient should be anticipated in part or entirely because of general inflation, the differences found in (a) would be reduced or eliminated.

14-24

For size 0 the annual cost of lost energy will go from $35.70 in year 1 to $107.10 in year 25. The gradient is $2.975. With an i of 14%, the equivalent uniform annual figure is
$35.70 + $2.975(6.161) = $35.70 + $18.33 = $54.03.

For size 1 the corresponding figures are $44.98, $134.94, and $3.74833. The equivalent annual figure now is
$44.98 + $3.74833(6.161) = $44.98 + $23.09 = $68.07

For size 2 the corresponding figures are $56.78, $170.34, and $4.73167. The equivalent annual figure now is
$56.78 + $4.73167(6.161) = $56.78 + $29.15 = $85.93.

14-25

If 3.4¢ is the unit cost at zero date and the increase is 0.68¢ per year, the unit cost for year 1 will be 4.08¢. If this gradient continues through year 10 and the unit cost then remains constant at 10.2¢ for the following 15 years, the equivalent uniform annual unit cost for the 25-year period may be computed as follows:

4.08¢ + [0.64¢(P/G,14%,10) + 10.2¢(P/A,14%,15)(P/F,14%,10)]·
(A/P,14%,25) = 4.08¢ + 4.13¢ = 8.21¢

With the foregoing equivalent annual unit cost, the equivalent annual costs of lost energy for wire sizes 0, 1, and 2 become $86.20, $108.61, and $137.11, respectively. When these are added to the respective figures in lines C and D of Table 14-7, the totals are $177.20, $180.78, and $194.45.

14-26

Make the comparison between size 0 and size 1 assuming that the change from 3.4¢ to 8.5¢ will take place by some selected date in the not too distant future. The comparison of equivalent uniform annual costs under this assumption might be made along the lines illustrated in the foregoing solution to Problem 14-25. By interpolation between the results obtained with an immediate change to a unit cost of 8.5¢ and such a change completed at the selected future date, an estimate can be made of how soon the change would need to be completed for annual costs for the two wire sizes to break even.

14-27

For reasons brought out in Example 14-3, during periods of inflation the prospective rate of return on an investment will be reduced where depreciation deductions to determine taxable income are to be based on the investment at zero date with no adjustment for changes in the purchasing power of the monetary unit. It follows that if an economy study is to be made in monetary units of constant purchasing power and if a value of i* has been stipulated, a recognition of this rule of income taxation will cause some increments of investment to be rejected that would be acceptable if the rule were not recognized. As applied to this choice among competing wire sizes, the forecast of inflation with no differential price change in the cost of energy would tend to favor lower investments and therefore <u>might</u> lead to selection of a smaller wire size than would otherwise be chosen.

14-28

(a) The investment tax credit at zero date will reduce the net cash flow at zero date by 10% for each wire size. Such a change reduces investment charges and thus tends to favor the larger investments. In this case it tends to favor the larger wire sizes.

(b) The more rapid depreciation for tax purposes tends to raise all prospective after-tax rates of return. Therefore, some larger wire size that would not otherwise be acceptable might now meet the criterion for acceptability.

(c) As brought out in Example 14-4, any such forecast tends to strengthen the case for a larger wire size.

(d) Because the copper wire must be purchased at year zero, the forecast of a differential change in its price should have no influence on the conclusions of the economy study as long as the change is not expected until a later date.

14-29

Estimated annual disbursements for items other than income taxes are $9,200 for Plan A and $5,100 for Plan B. The annual straight-line depreciation for income tax purposes is $15,000/10 = $1,500. The difference in estimated taxable income therefore is $9,200 - $5,100 - $1,500 = $2,600. With an applicable tax rate of 50%, this gives extra income taxes of 0.50($2,600) = $1,300 per year with Plan B.

14-29 continued

An analysis of the difference between Plans B and A along the lines of Problem 14-4 is as follows:
(Note that here Column F is -0.50E.)

Year	A	C	D	E	F	G	H
0	-$15,000	-$15,000				-$15,000	-$15,000
1	+4,100	+4,346	-$1,500	+$2,846	-$1,423	+2,923	+2,758
2	+4,100	+4,607	-1,500	+3,107	-1,554	+3,053	+2,717
3	+4,100	+4,803	-1,500	+3,383	-1,692	+3,191	+2,679
4	+4,100	+5,176	-1,500	+3,676	-1,838	+3,338	+2,644
5	+4,100	+5,487	-1,500	+3,987	-1,994	+3,493	+2,610
6	+4,100	+5,816	-1,500	+4,316	-2,158	+3,658	+2,579
7	+4,100	+6,165	-1,500	+4,665	-2,332	+3,833	+2,549
8	+4,100	+6,535	-1,500	+5,035	-2,518	+4,017	+2,520
9	+4,100	+6,927	-1,500	+5,427	-2,714	+4,213	+2,494
10	+4,100	+7,342	-1,500	+5,842	-2,921	+4,421	+2,469

The present worth of the series in Column H is zero with an i of approximately 11.8%. Therefore, the prospective rate of return is reduced from 13.3% to 11.8% by the prospect of 6% inflation combined with the expectation that inflation will not be recognized in computing depreciation for income tax purposes. The foregoing statement assumes that analysis is made in dollars of constant purchasing power.

14-30

Let n be the number of years of life to break even for the new material.

$$\$3,000(A/P,12\%,3) = \$5,800(A/P,12\%,n)$$

$$(A/P,12\%,n) = \frac{\$3,000}{\$5,800}(0.41635) = 0.21535$$

Interpolation between the capital recovery factors for 7 and 8 years (0.21912 and 0.20130, respectively) indicates that the new type of lining must last approximately 7.2 years to break even on annual cost.

14-31

The choice is between an extra \$90,000 (i.e., \$810,000 - \$720,000) now and an extra \$140,000 (i.e., \$540,000 - \$400,000) after n years. For the two to break even:

$$\$90,000(F/P,8\%,n) = \$140,000$$

$$(F/P,8\%,n) = 1.5556$$

Interpolation between the 8% compound amount factors for 5 and 6 years indicates that the new story ought to be needed within approximately 5.7 years for costs to break even at an i of 8%. In this problem the 40-year life is irrelevant because it does not influence any cash flow influenced by the choice between the alternatives.

14-32

This problem has more complications than may be evident at first glance.

Let A represent the design that requires an initial investment of \$720,000 and B the design that requires \$810,000. Annual property taxes with A will be 0.02(\$720,000) = \$14,400 for the first 5 years and 0.02(\$720,000 + \$540,000) = \$25,200 for the remaining 35 years. Annual property taxes with B will be 0.02(\$810,000) = \$16,200 for the first 5 years and 0.02(\$810,000 + \$400,000) = \$24,200 for the remaining 35 years.

14-32 continued

The choice between A and B will influence two types of deductions from taxable income throughout the 40-year life of the building, namely, the deductions for property taxes and the deductions for depreciation. The deductions affected will be as follows:

Years	A	B	B - A
1-5	\$14,400 + \$36,000 = \$50,400 per year	\$16,200 + \$40,500 = \$56,500 per year	+\$6,100 per year
6-20	\$25,200 + \$63,000 = \$88,200 per year	\$24,200 + \$60,500 = \$84,700 per year	-\$3,500 per year
21-25	\$25,200 + \$27,000 = \$52,200 per year	\$24,200 + \$20,000 = \$44,200 per year	-\$8,000 per year
26-40	\$25,200 per year	\$24,200 per year	-\$1,000 per year

Let t represent the applicable income tax rate over the 40-year life. With an i* of 8%, the present worth of the cash flow for (B - A) will be :

-\$90,000 + \$140,000(P/F,8%,5) - \$1,800(P/A,8%,5)
+\$1,000(P/A,8%,35)(P/F,8%,5) + \$6,100t(P/A,8%,5)
-\$3,500t(P/A,8%,15)(P/F,8%,5) -\$8,000t(P/A,8%,5)(P/F,8%,20)
- \$1,000t(P/A,8%,15)(P/F,8%,25)

With the appropriate compound interest factors inserted, the foregoing becomes:

- \$90,000 + \$140,000(0.6806)-\$1,800(3.993) + \$1,000(11.655)(0.6806)
+ \$6,100t(3.993) - \$3,500t(8.559)(0.6806) - \$8,000t(3.993)(0.2145)
-\$1,000t(8.559)(0.1460)

This reduces to +\$6,029 - \$4,133t

A positive net worth favors B. It is evident that with the stipulated after-tax i* of 8% this comparison favors B with any possible value of the applicable income tax rate. (The income tax rate would need to be 146% for the two alternatives to break even with an i* of 8%.)

14-33

(a) Sam's bracket now will be:

Over \$16,800 but not over \$20,160 . . .
\$2,735.25 plus 30% of excess over \$16,800

His tax will be \$2,735.25 + 0.30(\$17,600 - \$16,800)
= \$2,975.25.

His income tax has increased by \$2,975.25 - \$2,605
= \$370.25.
This is approximately 14.2% of \$2,605.

(b) Jennifer's bracket now will be:

Over \$31,290 but not over \$36,855 . . .
\$7,805.70 plus 44% of excess over \$31,290

Her tax will be \$7,805.70 + 0.44(\$33,000 - \$31,290)
= \$8,558.10.

Her income tax has increased by \$8,558.10 - \$7,522
= \$1,036.10.
This is approximately 13.8% of \$7,522.

14-34

If the burden of income taxation is not to be increased, taxes ought not to increase at a greater rate than the rate of inflation. In the preceding problem the combination of 10% inflation and "bracket creep" caused Sam and Jennifer to have tax increases of approximately 14% in spite of the 5% so-called tax reduction.

It would have been necessary to increase all dollar figures in Table 12-8 by multiplying them by 1.10 if it had been desired that tax burdens not be increased. With such a change Sam's bracket would have been:

Over \$15,290 but not over \$17,600 . . .
\$2,264.90 plus 26% of excess over \$15,290

His income tax would have been \$2,865.50. The increase of \$260.50 above the previous \$2,605 would have been exactly 10%.

Similarly, it may be calculated that Jennifer's tax would have increased from \$7,522 to \$8,274.20, an increase of \$752.20.

CHAPTER 15

Use of the Mathematics of Probability in Economy Studies

15-1

Expected annual cost of insurance plus fire damage.

	Sprinkler	No Sprinkler
Annual Ins. Premium	$12,960	$31,680
Expected Avg. Loss recovered (1/2 of above)	- 6,480	- 15,840
Expected Damage Suffered 3($2,400,000)(Ins. rate)(0.5)	19,440	47,520
	$25,920	$63,360

EUAC with Sprinkler system

= $85,600(A/P,30%,20) + $2,500 + $25,920

= $85,600(0.30159) + $2,500 + $25,920 = $54,236

Management might consider the possibility that a destructive fire, regardless of the insurance coverage, might cause the company to fail due to loss of customers, time required to rebuild, etc. The extreme event might be so frightening that a sprinkler system might be installed to reduce the probability of a fire of destructive proportions inspite of the outcome of the above analysis.

5-2

Insurance costs and expected values of loss are:

Annual costs:	Site A	Site B
Insurance premium	$9,750	$13,500
Expected loss recovery	-4,875	- 6,750
Expected loss	12,188	16,875
Total	$17,063	$23,625

Difference (B-A) = $6,562

The owner can afford to pay as much as $6,562/0.20 = $32,810 more for Site A than for Site B assuming that the full price paid for the land can be recovered at any time.

15-3

Design	First Cost	CR Cost	Risk Cost	Total Annual Cost
A	$200	$14,088	$40,000	$54,088
B	225	15,849	25,000	40,849
C	300	21,132	10,000	31,132
D	350	24,654	5,000	29,654
E	392	27,612	2,500	30,112
F	480	33,811	1,000	34,811

(A/P,7%,75) = 0.07044

Plan D, a spillway to handle 14,500 cfs, is the most economical

The solution is not sensitive to i in the range 5 to 9%, nor is it sensitive to a change of ± $100,000 in damages. Reducing the risk for Design E to from 0.005 to 0.003, however, shifts the economic choice to Design E.

A question occasionally arises in class as to whether this mode takes care of the probability of having a flood stage sufficient to exceed the spillway capacity more than once a year. That possibilit should be included in the statement of probabilities. In other words, the probability given is supposed to be for a year, and if there is a real probability of an excessive flow occurring more thar once, the probability should reflect it. The probability for a single event can be given for a month, a quarter, 6-month period, or a year. .If the damage is the same each time an overflow occurs, then the probabilities can be added for the different periods in the year in order to obtain a probability for the year.

15-4

(Using probabilities from Example 15-2)

Winning Spots	Probability	Payoff	Expected Value
0 to 3	0.938414048	$ 0	$0.000
4	0.052190967	6	0.313
5	0.008638505	120	1.037
6	0.000732077	2,450	1.794
7	0.000024403	25,000	0.610
Totals	1.000000000		$3.754

15-5

The buyer of a $6 7-spot ticket has an expected value of recovery of only 62.5% of the outlay for a ticket in contrast to the 75% obtained by the buyer of a 60-cent 7-spot ticket. The objective of the XYZ Club's "$25,000 rule" in its Keno games is to decrease the opportunity for a series of extreme events to ruin the XYZ Club or cause it severe financial trouble.

15-5 continued

Although the XYZ Club's rule making on Keno payoffs exists in quite a different setting from the design decisions mentioned in the problem statement, there is one element in common. Decision makers need to recognize that it sometimes is insufficient to consider merely expected values when decisions involve unlikely events that, if they occur, can have extremely adverse consequences to the persons or organizations for which the decision is made.

15-6

	Damages F(x)	Probability of stated damages, P(x)	E(x)
For Cap.	$ 0	(1.00 - 0.02)	$ 0
of 1900	250,000	(0.02 - 0.01)	2,500
cfs	300,000	(0.01 - 0.005)	1,500
	400,000	(0.005 - 0.000)	2,000
			$6,000
For Cap.	$ 0	(1.00 - 0.005)	$ 0
of 2,300	250,000	(0.005 - 0.002)	750
cfs	300,000	(0.002 - 0.001)	300
	400,000	(0.001 - 0.000)	400
			$1,450

15-7

(a) ($50,000/10)(P/A,15%,10) = $5,000(5.019)
= $25,095.

(b) ($50,000/10 - $50,000/50)(5.019) = $20,076.

15-8

Since two different amounts of damage may occur if the spillway capacity is exceeded, different costs of risk must be computed for each design. For example, the probability of a flow return between 8,500 and 10,000 cfs for Design A is (0.08-0.05) = 0.03; that for a flow greater than 10,000 cfs is 0.05. Thus the risk cost for Design A is $350,000(0.03)+$750,000(0.05) = $10,500+$37,500 = $48,000.

15-8 continued

Design	CR Cost	Risk Cost ($000 omitted)	Total
A	$14,088	$350(0.08-0.05)+$750(0.05)	$62,088
B	15,849	$350(0.05-0.03)+$750(0.03)	45,349
C	21,132	$350(0.02-0.01)+$750(0.01)	32,132
D	24,654	$350(0.01-0.0075)+$750(0.0075)	31,154
E	27,612	$350(0.005-0.002)+$750(0.002)	30,162
F	33,811	$350(0.002-0.001)+$750(0.001)	34,911

15-9

CR and Risk Costs for each alternative are shown in the solution to Problem 15-3.

Design	CR Cost	Risk Cost	Comparison	ΔCR	ΔRisk	B/C
A	$14,088	$40,000				
B	15,849	25,000	B vs A	$1,761	$15,000	8.52
C	21,132	10,000	C vs B	5,283	15,000	2.84
D	24,654	5,000	D vs C	3,522	5,000	1.42
E	27,612	2,500	E vs D	2,958	2,500	0.85
F	33,811	1,000	F vs D	9,157	4,000	0.44

The most economical choice is Design D.

15-10

Expected value of annual cost of damage to orange crop

$$= \$125{,}000(0.75)(4/50) = \$7{,}500$$

EUAC to prevent crop loss =

$$\$15{,}000(A/P,25\%,10) + \$2{,}000$$
$$= \$15{,}000(0.28007) + \$2{,}000 = \$6{,}201$$

The sensitivity of the solution to changes in the various factors can be tested by solving the same problem with several different values. An increase in the selling price of oranges will increase the expected value of the damage due to frosts. An increase in the minimum attractive rate of return will increase the equivalent uniform cost of providing protection against frosts. An increase in the probability of "killing" frosts will increase the expected value of loss due to frosts.

The owner's decision might be influenced by his financial resources and his ability to survive a killing frost (be able to meet mortgage payments, feed and house his family, etc.)

15-11

Expected annual cost = P(A/P,10%,50)
+ $30,000 (Probability of overtopping)

A. $32,500(0.10086) + $30,000(0.20) = $9,278
B. $4,236 + $3,000 = $7,236
C. $5,043 + $1,200 = $6,243
D. $6,253 + $600 = $6,853

15-12

Data from Problem 15-11.

Design	CR Cost	Risk Cost	Comparison	ΔCR	ΔB	B/C
A	$3,278	$6,000				
B	4,236	3,000	B vs. A	$958	$3,000	3.13
C	5,043	1,200	C vs. B	807	1,800	2.23
D	6,253	600	D vs. C	1,210	600	0.50

Design C is most economical choice.

15-13

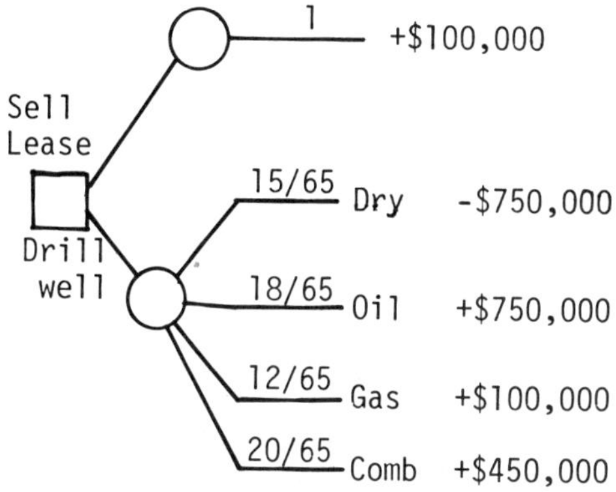

Selling yields a certain $100,000.
The expected net present value of drilling is $191,538.

15-14

NPW Payoff ($ x 1,000)

	Dry S_1	Oil S_2	Gas S_3	Comb. S_4	minimum
Drill A_1	-750	750	100	450	-750
Sell A_2	100	100	100	100	100

The maximum security level strategy is to sell the lease for \$100,000. If independent oil companies such as that described in Problem 15-13 consistently applied this decision rule, they would seldom do any drilling. Depending on the company's financial circumstances, it is likely to view the problem from an entirely different perspective when a site has a "track record" than when it does not. Previous successful drillings in comparable circumstances will affect the decision-makers intuition about the site.

15-15

$EPW(A_1)$ = \$140,000(0.6) + \$28,000(0.4) = \$95,200.

$EPW(A_2)$ = \$260,000(0.6) - \$80,000(0.4) = \$124,000.

$EPW(A_3)$ = [\$220,000(0.9) + \$20,000(0.1)](0.6) - \$20,000(0.4) = \$112,000.

15-16

(a) Minimums are: A_1, -36; A_2, -32; A_3, -35.

The choice should be A_2 with the maximum of the minimums, \$32,000 annual cost.

(b) $EAC(A_1)$ = -\$26(0.4)-\$30(0.2)-\$24(0.3)-\$36(0.1) = -\$27.2

$EAC(A_2)$ = -\$17(0.4)-\$32(0.2)-\$30(0.3)-\$28(0.1) = -\$25.0

$EAC(A_3)$ = -\$35(0.4)-\$20(0.2)-\$22(0.3)-\$30(0.1) = -\$27.6

A_2 is preferred because it has the lowest expected annual cost, \$25,000.

15-17

Present worth of costs ($000 omitted):

$3,675 + $475[(P/F,20%,5) + (P/F,20%,10)]

+ $150(P/A,20%,15) - $2,000(P/F,20%,15)

= $3,675 + $475(0.4019 + 0.1615) + $150(4.675)

- $2,000(0.0649) = $4,514.

Present worth of receipts:

$30(365)(150) = $1,642,500/yr. at 100% occupancy

PW($000 omitted) = $1,642.5(4.675) = $7,679.

ENPW = $7,679[0.8(0.3) + 0.6(0.6) + 0.35(0.1)]

- $4,514 = $362.

The company must always consider alternative uses for its funds. The present worth with only 35% occupancy is -$1,827,000 over the 15 years. They may wish to compare this figure with the low-end estimates of an alternative investment opportunity.

CHAPTER 16

Increment Costs and Sunk Costs

General Notes:

This chapter explains two concepts that are frequently violated by otherwise knowledgeable businessmen, and deserves considerable emphasis in the classroom.

The tendency to use averages is very strong, yet, as the very simple example of this chapter shows, can lead to very bad decisions The concept of incremental costs is very obvious when a little thought is given to it, yet accounting practices frequently lead engineers and businessmen to accept and use cost accounting data that include allocated costs that are irrelevant to the problem at hand. The question, "What are the actual differences that will occur with or without the proposed action?" will help to show the importance of understanding incremental cost analysis.

The question just previously suggested can also help the studen to understand the sunk cost concept. If he is forced to determine only those differences that lie in the future, he will quickly grasp the idea that a decision now cannot change things that occurre in the past, that a decision only affects the future.

Most of us have a psychological block against admitting that we have made a mistake, and frequently we try to cover up such mistakes by introducing some of the costs of those mistakes into the charges against future alternatives. This kind of error results from not understanding the sunk cost concept and the true situation one encounters when a decision is necessary.

16-1

Only the final 20 kw-hr of the 500 are priced at 8.1¢. The remaining 80 kw-hr of the proposed 100 kw-hr reduction are priced at 5.9¢. The reduction in the bill will be 20($0.081) + 80($0.059) = $1.62 + $4.72 = $6.34. When $6.34 is divided by 100 kw-hr the quotient is 6.34¢/kw-hr. Nevertheless, this average cost per kw-hr saved is not a unit cost that is significant for the purposes of any decision; the 8.1¢ and 5.9¢ figures are the relevant unit costs for decision making.

16-2

This calculated unit cost of 23.3¢ per kw-hr has no relevance to any decision because it should not be used as a multiplier for the number of kw-hr involved in any decision.

16-3

(1) (\$10,000 - \$5,300) + 0.15(\$10,000) = \$6,200

(2) (\$5,300 - \$2,900) + 0.15(\$5,300) = \$3,195

(3) (\$2,900 - \$2,100) + 0.15(\$2,900) = \$1,235

(4) (\$2,100 - \$1,600) + 0.15(\$2,100) = \$815

(5) (\$1,600 - \$1,200) + 0.15(\$1,600) = \$640

(6) (\$1,200 - \$900) + 0.15(\$1,200) = \$480

16-4

CR = (\$10,000 - \$900)(0.26424) + \$900(0.15) = \$2,540

or

PW of \$6,200	= \$6,200(0.8696)	=	\$5,392
PW of \$3,195	= \$3,195(0.7561)	=	2,416
PW of \$1,235	= \$1,235(0.6575)	=	812
PW of \$815	= \$815(0.5718)	=	466
PW of \$640	= \$640(0.4972)	=	318
PW of \$480	= \$480(0.4323)	=	208
Sum of the present worths		=	\$9,612

CR = \$9,612(0.26424) = \$2,540

16-5

CR = (\$5,300 - \$1,200)(0.35027) + \$1,200(0.15) = \$1,616

or

PW of \$3,195	= \$3,195(0.8696)	=	\$2,778
PW of \$1,235	= \$1,235(0.7561)	=	934
PW of \$815	= \$815(0.6575)	=	536
PW of \$640	= \$640(0.5718)	=	366
Sum of the present worths		=	\$4,614

CR = \$4,614(0.35027) = \$1,616

16-6

You have already spent $140. Nothing you can do now will change that. It is past and is a sunk cost. If you can assure a sale, and a commission income of $200, by spending some additional money, you will improve your situation if you can assure the sale by spending any amount less than $200. Thus, at this point in your decision process, the past expenditure at $140 is completely irrelevant except as it helps you predict the probability of success if additional money is spent.

16-7

If there were no income tax considerations, Frank and James ought to reason alike; a major element in the decision ought to be a forecast of the future price of XY stock. But it is possible that Frank might be inclined to borrow the $20,000 because of being reluctant to take a loss on his stock, and that James might be inclined to sell his stock because of being pleased at making a profit.

A sophisticated look at income tax considerations ought to make Frank more willing to sell his stock and to make James less willing. Frank would have a deductible capital loss that would reduce his current payment of income taxes. In contrast, James would have a taxable capital gain that would increase his current tax payments.

16-8

After each such sale and replacement of a unit the merchant had the same stock and $5 less cash. It is evident that he was $5 worse off after each such transaction regardless of what his accounts might have indicated.

16-9

Both friends are wrong. The alternatives are:

	Go to jail	Do not go to jail
Pay fine	--	-$20
Write story	+$25	--
	+$25	-$20

The difference between these two alternatives is $45.
(Note: Whose viewpoint and when the decision was made will be raised by some students. Did Steve know when he decided to go to jail rather than pay his fine that he could write and sell an article on the experience? The total difference of $45 is an after-the-fact analysis, and is the correct answer to the question raised in the problem, but it is not necessarily the basis for Steve's decision when he elected to go to jail.)

16-10

Demand charge:	
First 40 kw	$ 91.00
Next 240 kw @ $1.99	477.60
	$568.60
Energy charge:	
55,000 kw-hr @ 5.1¢	$2,805.00
Monthly bill	$3,373.60

Average cost = $3,373.60 ÷ 55,000 = 6.13¢/kw-hr

The proposed new machinery will increase the monthly bill as follows:

Demand charge:	
20 kw @ $1.99	$39.80
30 kw @ $1.82	54.60
	$94.40
Energy charge:	
2,000 kw-hr @ 5.1¢	$102.00
Addition to monthly bill	$196.40

Average cost of energy added is $196.40 ÷ 2,000 = 9.82¢/kw-hr

16-11

Reduction in demand charge = 30 @ $1.99	=	$59.70
Reduction in energy charge = 1,000 @ 5.1¢	=	51.00
Total reduction in bill		$110.70

This is an average saving of 11.07¢/kw-hr.

16-12

	Georgia (2700 T)	Florida (3600 T)
Total Cost: $115(2700)	$310,500	
$126(3600)		$453,600
Fixed Cost:	94,050	$162,000
Total Variable Cost:	$216,450	$291,600
Waste Paper Cost: $28(1440) +$36.50(0.8)(900)	66,600	
$30.50(0.8)(3,600)		87,840
Total Variable Cost other than W.P.	$149,850	$203,760
Per Ton	$ 55.50	$ 56.60

16-12 continued

In computing the waste paper cost for the Georgia plant, the 1440 tons available at $28 would produce 1440/0.8 = 1800 tons of finished paperboard. The remaining 900 tons of finished paperboard would be made from 900(0.8) = 720 tons of waste paper at $36.50 per ton.

Production scheduling in the short run will be done without regard to fixed costs.

Variable costs:

Georgia plant

$55.50 + $28(0.8) = $77.90/T up to 1800 T/mo. output

$55.50 + $36.50(0.8) = $84.70/T above 1800 T/mo. output

Florida plant

$56.60 + $30.50(0.8) = $81.00/T up to 4000 T/mo. output

$56.60 + $36.50(0.8) = $85.80/T above 4000 T/mo. output

The order of scheduling priority should be:

Georgia plant up to 1800 tons output
Florida plant up to 4,000 tons output
Georgia plant up to 3,600 tons output(capacity)
Florida plant up to 6,000 tons output(capacity)

in that order

(a) For a total production of 6,300 tons, 2300 tons should be scheduled for the Georgia plant and 4,000 tons for the Florida plant. This will result in a cost saving of $1,480 per month over present operating costs.

(b) For a total production of 9,000 tons:

Georgia	- 3,600	$386,730
Florida	- 5,400	$606,120
Total cost/mo.		$992,850

16-13

Both viewpoints are wrong. Jones and Smith now have the equivalent of $800, or each unit is worth $0.10. A more reasonable viewpoint, therefore, is that Jones or Smith should not spend a currency unit unless he believes he is getting $0.10 worth of goods for it. The values of the currency units before the change in rates are not relevant.

16-14

If it is assumed that an annual cost comparison is to be made for 9 years, the following criticisms are relevant.

1. Depreciation has been counted twice: once as a charge against continued ownership and again as a credit toward rental.
2. The charge for depreciation against continued ownership was based on the original cost when it should have been based on net realizable salvage value.
3. The $120 unusual repair cost was treated as an annual cost when it should be amortized over the remaining life (unless there is reason to suspect a recurrence of the uninsured damage.)

A correct cost analysis yields the following on a before-tax basis:

Cost of renting = $12(2)(12) = $288

EUAC of continued ownership for 9 years
($720 + $120)(A/P,i%,9) + $45 + $40
at 0% interest (which is equivalent to S.L. depreciation over remaining life) =
$840/9 + $85 = $178.33

@ 15%	$840(0.24036) + $85 = $286.90
@ 25%	$840(0.31634) + $85 = $350.73

The decision favors continued ownership at a before-tax RoR up to 19.2%; above that rate, renting is more economical.

16-15

Indirect manufacturing expense is made up of fixed costs plus indirect variable costs. Treating it as all variable cost gives incorrect effect to fixed costs. Since production scheduling assignments should always be based on variable cost analysis, consideration of total costs may lead to improper machine assignment at all levels of output. Manual and partly mechanized (general purpose) operations will have an improper advantage over highly mechanized or automated (special purpose) operations.

Hence, an attempt should be made to estimate "variable burden" (power, water, scrap, inspection, etc.) to be included with the costs of setup, direct labor, and materials for production scheduling purposes.

16-15 continued

The above comments apply for all levels of plant output up to plant capacity, including conditions (a) and (b). However, in the case of (c) the plant working at or near capacity with a large volume of unfilled orders, some modification of the above comments is appropriate. In considering the unfilled orders, the alternatives are:

I. Let the customers wait until plant capacity becomes available, and perhaps incur their wrath;

II. Process the backlog as soon as possible by:

(a) Purchasing additional equipment and hiring the needed labor, or
(b) Subcontracting out the work as possible.

In comparing the relative cost of these two alternatives, the next avoidable increment of cost is the total cost of new equipment or subcontracting, rather than just the variable cost of owned equipment.

16-16

It is unfortunate that the management writer confused the rate of return methodology per se with improper application. Were the analysis properly structured, the rate of return approach would be a useful guide to decision making.

Consider the following cash flow table of the alternatives available to John Doe, and the differences between these alternatives

Year	Buy Concession	Do Not Buy Concession	Difference
0	-$900	+$100	-$1000
1-3	+ 450		+ 450
3	+ 900		+ 900

Roe's offer decreased the relative attractiveness of Doe's initial offer (as it was intended to do), but the net superiority of the initial offer still probably remains attractive, yielding 42.8% as determined from:

$$0 = -\$1,000 + \$450(P/A,i\%,3) + \$900(P/F,i\%,3)$$

It is not proper to consider the rate of return for mutually exclusive alternatives separately, since the receipt of $100 for not buying the concession is dependent on the existence of the other alternative, and must be compared with it. That is, $100 will be received provided one foregoes the attractive cash flow of the concession.

16-17

(a)	San Francisquito	Los Trancos
Present Output	320 T.	400 T.
Avg. Cost/T.	$ 148	$ 160
Total Cost	$47,360	$64,000
Fixed Cost	12,480	23,200
Variable Cost	34,880	40,800
Var. Cost/T.	$ 109.00	$ 102.00

$7.00/T. can be saved by shifting production from S.F. to L.T. plant.

(b) The additional 90 T./mo. will cost $9,810/mo. if produced at S.F.; it will cost $9,180/mo. if produced at L.T. The monthly saving by producing at L.T. would be $630.00

16-18

Annual costs could be compared as follows:

Diode unit

CR = $125(A/P,20%,4)	=	$ 48
Extra operating costs	=	300
		$348

Paper-tape unit

CR = $795(A/P,20%,4) = $307

This before-tax comparison favors the paper-tape unit even with the assumption of a 4-year life.

16-19

(a) Demand charge:

First 5,000 kw	$9,700
Next 4,500 kw @ 1.67	7,515
	$17,215

Energy charge:

2,000,000 kw-hr @ 4.3¢	$86,000
Monthly bill	$103,215

Average cost is $103,215 ÷ 2,000,000 = 5.16¢/kw-hr

*The 4.3¢ price applies to all kw-hr up to 500,000 + 300(9,500) = 2,850,000 kw-hr. Because the actual consumption is less than this, all units purchased are priced at 4.3¢.

16-19 continued

(b) Extra demand charge will be:

500 kw @ \$1.67	\$835
500 kw @ \$1.48	740
	\$1,575

Extra energy charge will be:

350,000 kw-hr @ 4.3¢	\$15,050
Total added cost	\$16,625

The average cost of the extra energy will be \$16,625 ÷ 350,000 = 4.75¢/kw-hr

16-20

Demand charge still will be	\$17,215
Energy charge now is 1,200,000 kw-hr @ 4.3¢	51,600
Total monthly bill	\$68,815

Average cost is \$68,815 ÷ 1,200,000 = 5.73¢/kw-hr

16-21

It was calculated in the solution to Problem 16-19 that the energy charge per kw-hr would decline from 4.3¢ to 3.9¢ for all kw-hr purchased in excess of 2,850,000 per month. Because the maximum demand will be unchanged by the change to two-shift operation, this figure is still appropriate. The extra shift increases monthly energy consumption from 2,000,000 to 3,500,000 kw-hr. Of this extra 1,500,000 kw-hr, 850,000 will be purchased at 4.3¢ and the remaining 650,000 at 3.9¢.

850,000 @ 4.3¢	\$36,550
650,000 @ 3.9¢	25,350
Total	\$61,900

Because the demand charge will be unchanged, this \$61,900 will be the increase in the monthly electric bill. The average cost of the energy added by the second shift will be \$61,900 ÷ 1,500,000 = 4.13¢/kw-hr.

16-22

(1) (\$15,000 - \$10,000) + 0.18(\$15,000) = \$7,700

(2) (\$10,000 - \$6,000) + 0.18(\$10,000) = \$5,800

(3) (\$6,000 - \$3,000) + 0.18(\$6,000) = \$4,080

(4) (\$3,000 - \$1,000) + 0.18(\$3,000) = \$2,540

16-23

CR = (\$15,000 - \$1,000)(0.31174) + \$1,000(0.18) = \$5,384

or

PW of \$7,700 = \$7,700(0.8475)	=	\$6,526
PW of \$5,800 = \$5,800(0.7182)	=	4,166
PW of \$4,080 = \$4,080(0.6086)	=	2,483
PW of \$2,540 = \$2,540(0.5158)	=	1,310
Sum of the present worths	=	\$14,485

CR = \$14,485(0.37174) = \$5,385

The difference of \$1 in the two capital recovery costs is due to rounding the product to the nearest dollar in each multiplication.

CHAPTER 17

Economy Studies for Retirement & Replacement

17-1

	Decrease in Taxable Income	EUAC
Two year old oven:		
EUAC of foregone tax advantage due to loss disposal 0.5[$27,000 - 2($2,250)] (A/P,14%,10)		$2,157
Repairs & Maintenance	$1,500	1,500
Extra fuel cost	9,000	9,000
Depreciation	2,250	
	$12,750	$12,657
New oven:		
$58,000(A/P,14%,10) = $58,000(0.19171)		$11,119
Repairs and Maintenance	$2,000	2,000
Depreciation	5,800	
Extra income tax due to reduced deductions 0.5($12,750 - $7,800)		3,475
	$7,800	$16,594

17-2

	Decrease in Taxable Income	EUAC
Five year old equipment assuming 1-yr life:		
CR: 0.7($1,000)(A/P,12%,1)		$ 806
Extra labor & supplies	$3,000	3,000
Extra Income tax 0.5($4,838 - 3,000)		919
		$4,725
New Equipment:		
CR: [$22,500 - 0.1($22,500)] (A/P,12%,5)		$5,618
Equiv. Depreciation $22,500(5/15) - $1,500 (A/G,12%,5) =	$4,838	$5,618

17-2 continued

The old equipment should be retained even if only for one year. Should it be held for 2 or 3 years more, the only affect on the above analysis would be to reduce the CR cost and thereby make retention appear that much more favorable.

17-3

Assuming the costs as given in Problem 17-2, the equation is:

0.72($P)(A/P,12%,2) = $5,618 - $3,919
$P = $3,988

At any salvage value above $3,988, a decision on economy grounds would suggest replacement.

17-4

(a) Before-tax analysis

Defender EUAC = 3[$1,000(A/P,30%,5) -$500(A/F,30%,5) + $6,500] = $20,566

Challenger EUAC = $45,000(A/P,30%,5) -$15,000(A/F,30%,5) + $8,000 = $24,817

(b) After-tax analysis

Defender - BV now = 3($8,000/2) = $12,000
SV now = 3($1,000) = $ 3,000

Tax saving on loss if replaced now = 0.52($12,000 - $3,000) = $4,680

BV in 5 yrs = 3($8,000/4) = $6,000
SV in 5 yrs = 3($500) = $1,500

Tax saving on loss if replaced in 5 yrs = 0.52($6,000-$1,500) = $2,340

	Annual Tax Deductions	EUAC
CR = $7,680(A/P,15%,5)-$3,840(A/F,15%,5)		$1,722
Operating costs 3($6,500) =	$19,500	19,500
Depreciation	1,200	
	$20,700	$21,222

Challenger:

BV in 5 yrs = $45,000(15/55) = $12,273
SV in 5 yrs = $15,000

Tax cost on gain on disposal = 0.28($15,000-$12,273) = $764

CR = (1-0.1)$45,000(A/P,15%,5) -$14,236(A/F,15%,5)		$ 9,970
Operating Costs	$ 8,000	8,000
Equiv. depreciation $45,000(10/55)-$818(A/G,15%,5)	6,773	
Extra income tax 0.52($20,700-$14,773)		3,082
	$14,773	$21,052

17-4 continued

The before-tax analysis favors the defender; the challenger is favored by a small margin after taxes.

17-5

Old Flume:	EUAC
Maintenance	$6,800
Higher water loss:	
(0.15 - 0.03)(10M)($30.75/M)(365)	13,469
	$20,269

New Flume:	
CR = $185,000(A/P,8%,20) = $185,000(0.10185)	$18,842
Maintenance	2,000
	$20,842

The "Old Flume" analysis is a next-year cost; appropriate under the assumptions that the flume has no salvage value and will not leak in the future less than it will next year.

While it is most economical to keep the old flume for the present, the costs are so close that the irreducibles will likely swing the decision in favor of replacement. Possible future changes in the price of electricity also play an important part in the decision.

17-6

Plan A	EUAC
CR = $9,500(A/P,25%,10) = $9,500(0.28007)	$2,661
Maintenance	200
Energy (800 hrs.)(41 kw)($0.043/kw-hr)	1,410
	$4,271
Plan B	
CR = ($1,000+$5,100)(0.28007)	$1,708
Maintenance $250 + $175	425
Energy 2(1,200 hrs)(23kw)($0.043/kw-hr)	2,374
	$4,507

17-7

This requires only the recalculation of defender costs for Problem 17-1 using a 6-year life.

EUAC of foregone tax advantage = 0.5[$27,000-2($2,250)] (A/P,14%,6)-0.5[$27,000-8($2,250)](A/F,14%,6) = $2,893-$527 = $2,366.
Total defender EUAC = $2,366+$1,500+$9,000 = $12,866.
When this is compared with the challenger EUAC of $16,594, replacement should still not be made.

17-8

In reanalyzing Problem 17-4 (b), four cost elements will be affected.

(1) The challenger BV after 5 years will increase from $12,273 to $22,500. Thus, what was a "gain" on disposal will now be a "loss" with a tax savings of 0.52($22,500-$15,000) = $3,900.
(2) The CR cost will then become (1-0.1)$45,000(A/P,15%,5) -$18,900(A/F,15%,5) = $9,279 (instead of $9,978).
(3) The amount charged to depreciation will reduce from the equivalent $6,773 to ($45,000/10) = $4,500. Thus the tax deductibles reduce from $14,773 to $12,500.
(4) Extra income taxes will become 0.52($20,700-$12,500) = $4,264. Hence, the total EUAC for the challenger (after taxes) will become $9,279+$8,000+$4,264 = $21,543.

The defender (EUAC = $21,222) is now preferred by a slight margin.

17-9

P = $15,000, i = 20% before taxes.

t	S_t	A_t	MC_t	$\sum PW(MC_t)$	AC_t
0	$15,000				
1	-	-$24			
2	-	164			
3	5,800	930		$12,273	$5,828
4	4,400	1,680	$4,240	14,318	5,531
5	3,200	2,750	4,830	16,259	5,437
6	2,200	4,950	6,590	18,466	5,553

AC_3 = $15,000(A/P,20%,3) - $5,800(A/F,20%,3) +[-$24(P/F,20%,1) + $164(P/F,20%,2) + $930(P/F,20%,3)](A/P,20%,3) - $300 = $5,828. CAC = $5,437.

Defender next-year cost = MC_5 + $300 = $4,830+$300 = $5,130 = DAC.

Do not replace now.

17-10

Plan A	Decrease in Taxable income	EUAC
EUAC of dep.		
\$1727-\$173(A/G,13%,10)	\$1,120	
CR = \$9,500(1.0-0.1)(A/P,13%,10)		\$1,576
Maintenance	200	200
Energy	1,410	1,410
Extra income taxes = 0.5(\$3,667-\$2,730)		469
	\$2,730	\$3,655
Plan B		
EUAC of dep.		
\$927-\$92.7(A/G,13%,10)+\$267.	\$ 868	
CR = [\$1,000+\$5,100(1.0-0.1)] (A/P,13%,10)		1,030
Maintenance \$250+\$175	425	425
Energy	2,374	2,374
	\$3,667	\$3,829

Plan A is the most economical.

17-11

(a) If held for 10 years and sold for \$150,000, there will be a capital gain tax of 0.28(\$150,000-\$40,000) = \$30,800. The net recovery will therefore be \$119,200. The only figure affected is the CR cost of continued ownership which will be:

\$130,400(A/P,10%,10) - \$119,200(A/F,10%,10)
= \$130,400(0.16275) - \$119,200(0.06275) = \$13,743.

TotalEUAC for continued ownership is \$35,643, thus selling the warehouse is still more attractive.

(b) The indifference point occurs where:
\$130,400(A/P,10%,10)-[0.72(\$S)+0.28(\$40,000)](A/F,10%,10)
+ \$21,900 = \$35,300

from which:
0.04518(\$S) = \$7,120
\$S = \$157,592.

17-12

From Table 17-2, using i* = 15%

t	(n)	MC_t	$\sum PW(MC_t)$	AC_t
1	2	\$13,400	\$11,653	\$13,400
2	3	12,950	21,444	13,191
3	4	13,000	29,992	13,136
4	5	13,625	37,783	13,234

The defender advantage cost is \$13,136, the average cost for 3 years more service. While this may imply a remaining economic life of 3 years, that would yield a total economic life of only 4 years. While the DAC is \$13,136, this is for comparative purposes only against the CAC of \$15,143. The comparison simply means do not replace now. Replacement is made when the DAC exceeds the CAC.

17-13

From Table 17-2, the marginal costs are constantly increasing from year 4 on. Thus the defender advantage cost will be a next-year marginal cost each year after 3 year's service.

(a) DAC = MC_6 = \$14,825. Compared to CAC = \$15,143, replacement is not indicated now.

(b) DAC = MC_7 = \$16,025. Compared to CAC = \$15,143, replacement is now indicated.

Problems 17-12 and 17-13 illustrate that, if the proper figures are used as defender and challenger advantage costs, decisions in a steady-state environment will be entirely consistent.

17-14

Page 426 shows the annual cost for extending service 1 year more to be \$15,275 and the average annual cost for 3 years more to be \$15,308.

Thus

CC(1 yr. more for defender) = \$15,275(P/A,15%,1) + \$15,143(P/F,15%,1)/0.15 = \$101,072.

CC(3 yrs more for defender) = \$15,308(P/A,15%,3) + \$15,143(P/F,15%,3)/0.15 = \$101,325.

Each of these capitalized cost figures assumes that the defender, when replaced, will be replaced by the same challenger and all of its successors. These successors are assumed to have the identical cash flow time profiles to the present challenger.

17-15

Four Year Life:	Annual Cost
CR = \$46,000(A/P,18%,4)-\$20,000(A/F,18%,4)	
= \$46,000(0.37174)-\$20,000(0.19174) =	\$13,265
Disbursements = \$6,000+\$500(A/G,18%,4)	
= \$6,000 + \$500(1.295) =	6,648
	\$19,913
Eight Year Life:	
CR = \$46,000(A/P,18%,8)-\$10,000(A/F,18%,8)	
= \$46,000(0.24524)-\$10,000(0.06524) =	\$10,629
Overhaul = \$8,500(P/F,18%,4)(A/P,18%,8)	
= \$8,500(0.5158)(0.24524) =	1,075
Disbursements = \$6,000+\$500(A/G,18%,8)	
= \$6,000+\$500(2.656) =	7,328
	\$19,032
Twelve Year Life:	
CR = \$46,000(A/P,18%,12)-\$5,000(A/F,18%,12)	
= \$46,000(0.20863)-\$5,000(0.02863) =	\$ 9,454
Overhaul = [\$8,500(P/F,18%,4)+\$10,000(P/F,18%,8)] (A/P,18%,12)=[\$8,500(0.5158)+\$10,000(0.2660)] (0.20863) =	1,470
Disbursements = [\$6,000(P/A,18%,12)+\$500(P/G,18%,8) +\$7,000(P/A,18%,4)(P/F,18%,8)](A/P,18%,12)	
= [\$6,000(4.793)+\$500(10.829) + \$7,000(2.690)(0.2660)](0.20863) =	8,174
	\$19,098

17-16

New tractor (the challenger) for eight year life:
The CAC = \$19,032 (from Problem 17-15).

8-yr-old tractor (the defender):	
CR = (\$10,000+\$10,000)(A/P,18%,4) -\$5,000(A/F,18%,4) =	\$ 6,476
Annual disbursements =	13,000
	\$19,476

Make the replacement now.

17-17

New tractor (the challenger) for 8 year life:
CAC = \$19,032 (from Problem 17-15).

4-year-old tractor (the defender):

CR = (\$20,000+\$8,500)(A/P,18%,4)-\$10,000(A/F,18%,4) = \$8,677
Annual disbursements = \$8,000+\$500(A/G,18%,4)
= \$8,000+\$500(1.295) = 8,648
\$17,325

Retain the defender.

In order for replacement to be attractive, the present salvage must be at least:

\$P(A/P,18%,4) = \$19,032 - \$8,648 - \$1,242
from which \$P = \$24,593.

17-18

(a) Annual disbursements have averaged: 1-\$175; 2-\$224; 3-\$336; 4-\$511; 5-\$735; 6-\$1,106. If this trend continues, it should not be replaced now because the next-year cost will probably be about \$1,512, which is less than the minimum EUAC for an average replacement (\$2,048).
(b) Since the last-year disbursements (\$2,120) exceed the average minimum EUAC (\$2,048), the machine should be replaced despite the fact it is only 5 years old.
(c) Since a general inflation increases all costs proportionately, the economic time of replacement will remain the same. All of the costs in the table will simply be increased by a factor of 1.3.
(d) If it is assumed that all four existing machines are 5 years old, presumably they would have one year of economic life remaining. However, since the four replacements can be purchased for \$900 each, \$700 less than presumably the open market price of \$1600, replacement now (losing one year on each existing machine) will buy four years of economic service. Balanced against an average cost of \$2,048 for 6 years normal service, these replacement machines will have an average cost of:
\$900(A/P,20%,4) + [\$480(P/F,20%,1) + \$730(P/F,20%,2) + \$1,050(P/F,20%,3) + \$1,580(P/F,20%,4)](A/P,20%,4) = \$1,199.

17-19

Challenger:

CR = $27,500(A/P,20%,10) =	$6,559
Labor & extras = $400(48)(1.30) =	24,960
Power & Maintenance = $0.05(4,000)+$600 =	800
Taxes and Insurance 0.03($66,000) =	1,980
Total annual cost	$34,299

Defenders:

CR = 2($2,000)(A/P,20%,10) =	$ 954
Labor & Extras = $250(2)(1.3)(48) =	31,200
Power & Maintenance = 2[2,500($0.05)+$400]=	1,050
Taxes & insurance 2(0.03)($15,000) =	900
	$34,104

The difference in annual cost between these two alternatives is so small that the decision is likely to be made on factors not reduced to money terms.

17-20

Challenger (Same costs itemized in Problem 17-19.)

	Decrease in Taxable Income	EUAC
CR = (1-0.1)$27,500(A/P,10%,10) =		$4,028
Annual disbursements	$27,740	27,740
Equiv. ann. dep. = $5,000-$500(A/G,10%,10)=	3,138	
Extra annual income taxes 0.5($33,150-$30,878)=		1,136
Totals	$30,878	$32,904

Defenders

CR = 2[0.5($5,333-$2,000)+$2,000](A/P,10%,10)=		$ 1,193
Annual disbursements =	$33,150	33,150
Totals	$33,150	$34,343

Challenger clearly is preferred in this after-tax analysis.

17-21

The difference in annual cost between the challenger and defender in Problem 17-20 is $1,439 in favor of the challenger. Assuming that each future year's challenger will offer a $150 saving over the previous challenger, the annual cost of this for 20 years is:

$$\$150(A/G,10\%,20) = \$150(5.365) = \$805$$

In perpetuity it is:

$$\$150(A/G,10\%,\infty) = \$150(6.667) = \$1,000$$

Since neither of these figures exceeds the annual savings expected from this year's challenger, the decision still favors replacement now.

17-22

(a)

Year	Plan A	Plan B	(B-A)
0	-$4,500-$24,000 = -$28,500	-$36,000	-$7,500
1-5	-$14,100-$12,375 = -$26,475	- 21,700	+ 4,775
5	+$3,000+$7,150 = +$10,150	+ 13,750	+ 3,600

$0 = -\$7,500+\$4,775(P/A,i\%,5)+\$3,600(P/F,i\%,5)$; $i>50\%$

(actually 60.4%)

(b) After-tax analysis

Plan A

XY-10 current book value =(10/15)$15,000 = $10,000; loss on disposal if sold now = $10,000-$4,500 = $5,500*

XY-10 book value 5 years hence = (5/15)$15,000 = $5,000; Loss on disposal = $5,000-$3,000 = $2,000*

XY-12 book value 5 years hence = (55/120)$24,000; = $11,000; loss on disposal = $11,000-$7,150 = $3,850*

Year	CFBT	Dep. deduct	Taxable	Tax	CFAT
0	-$28,500		-$5,500*	+$2,750	-$25,750
1	- 26,475	-$1000-$3,000= -$4,000	-30,475	+15,238	- 11,237.5
2	- 26,475	- 1000- 2,800= - 3,800	-30,275	+15,137	- 11,337.5
3	- 26,475	- 1000- 2,600= - 3,600	-30,075	+15,038	- 11,437.5
4	- 26,475	- 1000- 2,400= - 3,400	-29,875	+14,937	- 11,537.5
5	- 26,475	- 1000- 2,200= - 3,200	-29,675	+14,838	- 11,637.5
5	+ 10,150		- 5,850*	+ 2,925	+ 13,075

17-22 Continued

Plan B XY-20 book value 5 years hence = (55/120)\$36,000 = \$16,500; loss on disposal = \$16,500-\$13,750 = \$2,750*.

Year	CFBT	Dep. Deduct.	Taxable	Tax	CFAT	(B-A)
0	-\$36,000				-\$36,000	-\$10,250
1	- 21,700	-\$4,500	-\$26,200	+\$13,100	- 8,600	+ 2,637.5
2	- 21,700	- 4,200	- 25,900	+ 12,950	- 8,750	+ 2,587.5
3	- 21,700	- 3,900	- 25,600	+ 12,800	- 8,900	+ 2,537.5
4	- 21,700	- 3,600	- 25,300	+ 12,650	- 9,050	+ 2,487.5
5	- 21,700	- 3,300	- 25,000	+ 12,500	- 9,200	+ 2,437.5
5	+ 13,750		- 2,750*	+ 1,375	+ 15,125	+ 2,050

0 = -\$10,250+\$2,637.5(P/A,i%,5)-\$50(P/G,i%,5)+\$2,050(P/F,i%,5);
i = 12.4%. Note that this is much less than half of the before-tax return.

(c) These values of i should be interpreted as the rate of return on the incremental investment resulting from implementing Plan B rather than Plan A. If 12.4% after income taxes is acceptable, Plan B is preferred.

17-23

(a) The total investment in purchase and installation of Machine X is \$120,000+\$24,000 = \$144,000. The total savings for 8 years is \$32,000+\$25,000+\$22,000+\$19,000+\$16,000+\$13,000 +\$10,000+\$7,000 = \$144,000. Therefore, the rate of return on total investment is 0%.

(b) This rate of return is not relevant to the decision at hand. The \$120,000 is a sunk cost, and only that portion of it, the present market value of \$36,000, should properly be considered in addition to the installation cost of \$24,000.

(c) If Alternative I is chosen, there is no investment. If Alternative II is chosen, there is an investment of the market value of Machine X, or \$36,000, plus the installation cost of \$24,000, for a total of \$60,000. Therefore the increased investment of Alternative II is \$60,000.

End of Year	Before-Tax Cash Flow Difference Alternative II less Alternative I
1977:0	-\$60,000
1	+ 32,000
2	+ 25,000
3	+ 22,000
4	+ 19,000
5	+ 16,000
6	+ 13,000
7	+ 10,000
1985:8	+ 7,000

17-23 continued

(d) Before-tax rate of return:

$$0 = -\$60{,}000 + (\$32{,}000-\$28{,}000)(P/F,i\%,1) + \$28{,}000(P/A,i\%,8) - \$3{,}000(P/G,i\%,8)$$

Which gives an i = 34.7%.

(e) Difference in Cash Flow, Alternative II - Alternative I (Headings same as in Problem 17-22 (b), Plan A.)

0	$-60,000		+$84,000*	-$42,000	-$102,000
1	+32,000	-$32,000	0	0	+ 32,000
2	+25,000	- 28,000	- 3,000	+ 1,500	+ 26,500
3	+22,000	- 24,000	- 2,000	+ 1,000	+ 23,000
4	+19,000	- 20,000	- 1,000	+ 500	+ 19,500
5	+16,000	- 16,000	0	0	+ 16,000
6	+13,000	- 12,000	+ 1,000	- 500	+ 12,500
7	+10,000	- 8,000	+ 2,000	- 1,000	+ 9,000
8	+ 7,000	- 4,000	+ 3,000	- 1,500	+ 5,500

*Foregone "loss" on disposal = ($120,000-$36,000) = +$84,000. The above solution also assumes that the $24,000 installation cost was capitalized and depreciated along with the $120,000 cost of Machine X.

(f) The rate of return on the extra investment in Alternative II may be found from the relationship:

$$0 = -\$102{,}000+\$32{,}000(P/F,i\%,1) + \$26{,}500(P/F,i\%,2) + \$23{,}000(P/F,i\%,3) + \ldots + \$5500(P/F,i\%,8)$$

i = 11.3%

(g) If a rate of return on capital investments of about 11.3% is satisfactory to management then Alternative II should be approved. This rate of return reflects the differences between the alternatives starting from now (November, 1977) and is the only rate of all those computed above that is pertinent to a realistic decision.
It is interesting to note the substantial disadvantage given Alternative II in the after-tax analysis.

CHAPTER 18

The Influence on Economy Studies of Sources of Investment Funds

General Notes:

A great many mistakes are made in business decisions due to the use of the prospective rate of return on the equity portion of a proposed investment as a measure of the attractiveness of the proposal. The decision should be based on the prospective rate of return on the proposed project as if it were to be entirely financed from equity funds. That is the only true measure of the attractiveness of the investments.

When a portion of an investment is to be financed by borrowed money at an interest rate lower than the i earned by the project, the RoR on equity will be higher than i. As the ratio of borrowed to equity funds increases, the RoR on the equity will approach infinity as the equity portion decreases to zero. Obviously, measuring the attractiveness of a proposal by the RoR on equity is like using a rubber tape measure. It is not reliable.

A second major item in Chapter 18 is the matter of leasing vs. ownership. This chapter clearly shows that leasing is a form of borrowing or renting. A person who realizes that costs of leasing and borrowing can be measured in the same way is in a favorable position to evaluate the sales arguments of leasing agents. Leasing is justified in many situations, but the decision to lease is too frequently made for the wrong reasons.

The teacher can bring the student to understand leasing by using examples, such as the leasing of private autos, for which the student can obtain information from local agents. The experience will usually make leasing more understandable and "real" to the student.

By this point in the course, the student should be able to handle some rather difficult analysis problems. This chapter gives the teacher the opportunity to find out how well the student has learned the basic principles.

18-1

End of Year	Plan P	Plan R	Plan P - Plan R
0	-100N	-10N - 24N	-66N
1		-24N	+24N
2		-24N	+24N
3	+ 55N	+10N	+45N

The interest rate that equates the two plans is found from:

PW = 0 = -66N + 24N(P/A,i%,2) + 45N(P/F,i%,3); i = 17.1%

18-2

Deductions from taxable income with Plan P will be 22.5N, 15N, and 7.5N for years 1, 2, and 3, respectively. Deductions with Plan R will be 24N for each of these three years. Differences in deductions will be 1.5N, 9N, and 16.5N. With an applicable tax rate of 45%, Plan R will have lower income taxes of 0.675N, 4.05N, and 7.425N. When these tax differences are combined with the differences in before-tax cash flow found in the solution to Problem 18-1, the equation to find the after-tax cost of the money made available by leasing is:

0 = -66N + (24N - 0.675N)(P/F,i%,1) + (24N - 4.05N)(P/F,i%,2)
+ (45N - 7.425N)(P/F,i%,3)

i = approximately 9.9%.

18-3

Consider first the before-tax analysis in the solution to Problem 18-1. Cash flow for Plan R will be unchanged whereas the positive cash flow for Plan P in year 3 will be reduced below 55N. Therefore the difference in cash flow in year 3 will be less than the +45N indicated in the solution. The before-tax cost of money will be reduced. The expectation of a lower salvage value makes leasing more attractive to a prospective lessee. This answer assumes that the lessor's price is not influenced by the lessee's expectation about salvage value.

The foregoing statement also applies to an after-tax analysis provided the lessee's estimate does not influence either the allowable depreciation deduction from taxable income or the required payments under the lease. If one or both of these will be affected, a general answer is not possible; each case calls for its own analysis.

18-4

The before-tax cost of the money provided by the lessor in Problem 18-1 was 17.1%; if the applicable tax rate had been 0%, this would also have been the after-tax cost of the money. With an applicable tax rate of 45% in Problem 18-2, this after-tax cost was reduced to 9.9%. If the tax rate had been 50% in Problem 18-2, the cost would have been 9.1%. In general, high applicable income tax rates tend to encourage financing by leasing rather than ownership just as they tend to encourage financing by borrowing rather than ownership. The extent of the encouragement depends on the rules established by the taxing authorities about rates of write-off if assets are owned and about the deductibility of payments to the lessor if assets are leased.

18-5

(a) 0 = -\$60,000 + \$19,400(P/A,i%,5)

i = approximately <u>18.5%</u>

(b) Debt charge the first year will be \$10,000 + (0.12)(\$50,000) = \$16,000. This will decrease by (0.12)(\$10,000) = \$1,200 each year. The positive cash flow in year 1 will be \$19,400 - \$16,000 = \$3,400. This will increase by \$1,200 each year.

0 = -\$10,000 + \$3,400(P/A,i%,5) + \$1,200(P/G,i%,5)

i = approximately <u>41.3%</u>

18-6

(a) Depreciation will be 5/15(\$60,000) = \$20,000 in year 1 and will decrease by 1/15(\$60,000) = \$4,000 each year. Taxable income will be \$19,400 - \$20,000 = -\$600 in year 1 and will increase by \$4,000 each year through year 5. The after-tax cash flow will be \$19,400 - (0.45)(-\$600) = + \$19,670 in year 1 and will decrease by (0.45)(\$4,000) = \$1,800 each year

0 = -\$60,000 + \$19,670(P/A,i%,5) - \$1,800(P/G,i%,5)

i = approximately <u>11.6%</u>

18-6 continued

(b) Before-tax cash flow is taken from the solution to 18-5(b)

Year	Before-tax cash flow	Taxable Income	Cash Flow for taxes	After-tax cash flow
0	-$10,000			-$10,000
1	+3,400	$19,400-$20,000-$6,000 = -$6,600	+$2,970	+6,370
2	+4,600	19,400- 16,000- 4,800 = -1,400	+630	+5,230
3	+5,800	19,400- 12,000- 3,600 = +3,800	-1,710	+4,090
4	+7,000	19,400- 8,000- 2,400 = +9,000	-4,050	+2,950
5	+8,200	19,400- 4,000- 1,200 = +14,200	-6,390	+1,810

0 = -$10,000 + $6,370(P/A,i%,5) - $1,140(P/G,i%,5)

i = approximately 38.7%

18-7

The implication in this problem is that the payments under the lease are fixed in dollars (or other relevant currency units) regardless of the rate of inflation. In contrast, presumably inflation will increase the resale value (55N). If an analysis is made along the lines indicated in Chapter 14 using monetary units that have the purchasing power existing at zero date, and if the rate of inflation is 4%, the series from the solution to Problem 18-1 will become:

Year	Plan P - Plan R
0	-66N
1	+24N/1.04 = 23.076N
2	$+24N/(1.04)^2 = 22.190N$
3	$+55N(1.04)^3/(1.04)^3 - 10N/(1.04)^3 = 46.11N$

0 = -66N + 23.076N(P/F,i%,1) + 22.19N(P/F,i%,2) + 46.11N(P/F,i%,3)

i = approximately 15.9%. This calculated cost of money is based on using currency units of constant purchasing power. It is slightly lower than the 17.1% figure, calculated in the solution to Problem 18-1, which would have measured the cost of money if no inflation had been anticipated. Therefore the prospect of inflation favors leasing in the sense that leasing appears slightly less unfavorable. It should be noted that for these particular facts and estimates leasing will still be very costly even with inflation.

18-8

The combination of the solutions of Problems 18-2 and 18-7 gives the following series of estimated cash flows expressed in dollars that have the purchasing power of dollars at zero date:

Year	Plan P - Plan R
0	$- 66N$
1	$+23.076N - 0.675N/1.04 = 22.427N$
2	$+22.190N - 4.05N/(1.04)^2 = 18.445N$
3	$+46.11N - 7.425N/(1.04)^3 = 39.509N$

$$0 = -66N + 22.427N(P/F,i\%,1) + 18.445N(P/F,i\%,2) + 39.509(P/F,i\%,3)$$

i = approximately 9.5%

This cost of funds based on currency units of constant purchasing power is slightly lower than the 9.9% computed in the solution to Problem 18-2. Therefore the prospect of inflation will cause a small reduction in the cost of financing by leasing if an analysis is made along the lines suggested in Chapter 14.

18-9

(a) $0 \doteq -\$90{,}000 + \$22{,}000(P/A,i\%,9)$

i = approximately 19.6% before taxes on the total investment.

(b) Debt charges the first year will be \$7,000 +(\$63,000)(0.12) = \$14,560. These charges will decrease by (\$7,000)(0.12) = \$840 each year. The positive cash flow in year 1 will be \$22,000 - \$14,560 = \$7,440. This will increase by \$840 each year.

$$0 = -\$27{,}000 + \$7{,}440(P/A,i\%,9) + \$840(P/G,i\%,9)$$

i = approximately 32.0% after taxes on the equity portion of the investment.

18-10

(a) Depreciation will be 9/45(\$90,000) = \$18,000 in year 1 and will decrease by 1/45(\$90,000) = \$2,000 each year. Taxable income will be \$22,000-\$18,000 = +\$4,000 in year 1 and will increase by \$2,000 each year. The after-tax cash flow will be \$22,000 - (0.45)(\$4,000) = \$20,200 in year 1 and will decrease by (0.45)(\$2,000) = \$900 each year.

$$0 = -\$90{,}000 + \$20{,}200(P/A,i\%,9) - \$900(P/G,i\%,9)$$

i = approximately 12.6% after taxes on the total investment

(b) Before-tax cash flow is taken from the solution to 18-9(b).

18-10 continued

Year	Before-tax cash flow	Taxable income		Cash flow for taxes	After-tax cash flow
0	-$27,000				-$27,000
1	+7,440	$22,000-$18,000-$7,560 =	-$3,560	+$1,602	+9,042
2	+8,280	22,000- 16,000- 6,720 =	-720	+324	+8,604
3	+9,120	22,000- 14,000- 5,880 =	+2,120	-954	+8,166
4	+9,960	22,000- 12,000- 5,040 =	+4,960	-2,232	+7,728
5	+10,800	22,000- 10,000- 4,200 =	+7,800	-3,510	+7,290
6	+11,640	22,000- 8,000- 3,360 =	+10,640	-4,788	+6,852
7	+12,480	22,000- 6,000- 2,520 =	+13,480	-6,066	+6,414
8	+13,320	22,000- 4,000- 1,680 =	+16,320	-7,344	+5,976
9	+14,160	22,000- 2,000- 840 =	+19,160	-8,622	+5,538

0 = -$27,000 + $9,042(P/A,i%,9) - $438(P/G,i%,9)

i = approximately 25.6% after taxes on the equity portion of the investment.

18-11

Inflation causes a differential price change favorable to the borrower and unfavorable to the lender, as was pointed out in Chapter 14. Therefore the prospect of inflation favors (b), financing by a combination of debt and equity.

18-12

Year	Before-tax cash flow	Write-off	Taxable income	Cash flow for taxes	After-tax cash flow
0	-$4,000				-$4,000
1	+ 2,600	-$2,500	+$2,100	-$1,050	+ 1,550
2	+ 2,800	- 2,500	+ 2,300	- 1,150	+ 1,650
3	+ 3,000	- 2,500	+ 2,500	- 1,250	+ 1,750
4	+ 3,200	- 2,500	+ 2,700	- 1,350	+ 1,850
5	+ 3,400	- 2,500	+ 2,900	- 1,450	+ 1,950
6	+ 3,600	- 2,500	+ 3,100	- 1,550	+ 2,050
7	+ 3,800	- 2,500	+ 3,300	- 1,650	+ 2,150
8	+ 4,000	- 2,500	+ 3,500	- 1,750	+ 2,250

0 = -$4,000 + $1,550(P/A,i%,8) + 100(P/G,i%,8)

i = 40.7%

18-13

Assume the first cost of the hydroelectric power project is P and the life is n (less than 50 years). Then the costs the magazine writer would include are:

(1) P(i)
(2) P/n
(3) P(A/F,i%,n)
(4) P(A/P,i%,n) or its non-uniform equivalent
(5) Operation and maintenance costs

Thus capital recovery is provided for thrice, and interest on capital recovery twice. Either (1) and (3), or (4) alone, would be adequate to provide for capital recovery with interest, since:

$$P(i) + P(A/F,i\%,n) = P(A/P,i\%,n)$$

(2) alone would provide for just capital recovery without recognizing the time value of money.

Operation and maintenance costs (5) should be included with any one of the above provisions for capital recovery with interest to arrive at total annual cost.

18-14

Total expenses, as tabulated from the garage owner's association report, are $445,950 per year.

Calculated at 7% interest, the EUAC to the city is ($3,200,000 - $700,000)(A/P,7%,50) + $700,000(0.07) = $181,150 + $49,000 = $230,150 This is less than half the cost quoted above.

1. The depreciation charge assumes recovery of the full cost of the construction of the structure over 50 years at 0% interest. (A/P,0%,50) = 1/50.

2. Bond recovery with interest assumes payment of all costs, building and land, over 20 years. The interest charge of $224,000 is the first year charge only. It will decline by 0.07($160,000) = $11,200 annually after the first year.

3. The sinking fund at 5% provides for recovery of the full capital investment over 50 years. This all adds up to triple counting of various investment costs over the period.

It has been assumed that the structure will have 0 salvage at the end of 50 years. If revenues were set such that the debt obligation of the city were covered ($160,000 + $224,000 = $384,000), the garage would be self-sustaining and would more than pay for itself. Another option, to reduce annual costs, would be to lease air rights above the garage.

18-15

(a) Because of an improper choice of alternatives to compare, Mr. A's analysis is misleading. However, he has correctly computed the differences in cash flow after taxes for the lease vs. purchase situation. In addition, his calculation of the compound amount ten years hence is correct, assuming that funds made available through leasing will earn 20% net after taxes in other available alternative uses on a risk-adjusted basis. This may be verified as follows: The PW of the difference in cash flow after taxes at 20% is:

$$\begin{aligned}&\$100{,}000 - \$25{,}000(P/A,20\%,5) + \$1{,}000(P/G,20\%,5)\\ &\quad -[\$8{,}000(P/A,20\%,4) - \$1{,}000(P/G,20\%,4)](P/F,20\%,5)\\ &\quad + \$6{,}000(P/F,20\%,10)\\ &= \$24{,}106\end{aligned}$$

Then the compound amount in 10 years is \$24,106(F/P,20%,10) = \$149,265, substantially the same as Mr. A's \$149,300.

(b) The before-tax cost of money obtained from leasing:

$$\begin{aligned}0 &= +\$100{,}000-\$30{,}000(P/A,i\%,5) - \$6{,}000(P/A,i\%,4)(P/F,i\%,5)\\ &\quad + \$4{,}000(P/F,i\%,10);\quad \underline{i = 18.2\%}.\end{aligned}$$

(c) The after-tax cost of money obtained from leasing:

$$\begin{aligned}0 &= +\$100{,}000-\$25{,}000(P/A,i\%,5) + \$1{,}000(P/G,i\%,5)\\ &\quad -[\$8{,}000(P/A,i\%,4) - \$1{,}000(P/G,i\%,4)](P/F,i\%,5)\\ &\quad + \$6{,}000(P/F,i\%,10);\quad \underline{i = 9.5\%}.\end{aligned}$$

(d) It is not relevant to use the compound amount at some future date as a decision criteria concerning present decisions.

Mr. A. is assuming that if the \$100,000 were not available from leasing to earn the 20%, that these earnings would be foregone; that is, that there are no other sources of investment funds available to the company. The analysis would be properly structured only, however, if he considered either:

(i) the difference between earnings from funds made available from leasing and funds made available from a bank line of credit or other source, since other sources are normally available; or

(ii) omitted completely the consideration of working capital earnings, since they would be the same regardless of their source. What does vary is the cost of money from the various sources.

It would be better if he chose (ii), but if he feels he must give effect to what happens to the freed funds, he may choose (i). Procedure (ii) is usually followed in economy studies to simplify the analysis.

18-15 continued

Leasing would be preferable if the after-tax cost of money from leasing, 9-1/2%, were less than the cost of funds available from other sources after taxes (which it probably is not). Thus it would be necessary to know either the cost of other funds or some statement of the after-tax minimum attractive rate of return for projects of this nature.

18-16

Consider first the differences in disbursements before income taxes. At zero date rental requires a disbursement of 18C as a deposit plus 20C for the first year's rental, a total of 38C. This is in contrast to a disbursement of 100C for purchase. Subsequent cash flows under rental are -20C for years 1, 2, and 3; -16C for years 4, 5, and 6; -10C for years 7, 8, and 9; and +18C for year 10.

With rental, each rental payment is a deduction from taxable income for the year for which the rental is prepaid. For example, the rental of 20C paid at zero date reduces taxable income for year 1. The deposit of 18C at zero date and its recovery at date 10 have no influence on taxable income. Under ownership, years-digits depreciation is a deduction from taxable income. For each year there will be a greater deduction from taxable income with leasing than with ownership. The differences in cash flow before and after taxes will be as follows:

Year	Before-tax difference	Difference in Taxes	After-tax difference
0	+62C		+62.00C
1	-20C	0.45(20C - 18.18C) = +0.82C	-19.18C
2	-20C	0.45(20C - 16.36C) = +1.64C	-18.36C
3	-20C	0.45(20C - 14.55C) = +2.45C	-17.55C
4	-16C	0.45(20C - 12.73C) = +3.27C	-12.73C
5	-16C	0.45(16C - 10.91C) = +2.29C	-13.71C
6	-16C	0.45(16C - 9.09C) = +3.11C	-12.89C
7	-10C	0.45(16C - 7.27C) = +3.93C	-6.07C
8	-10C	0.45(10C - 6.45C) = +1.60C	-8.40C
9	-10C	0.45(10C - 3.64C) = +2.86C	-7.14C
10	+18C	0.45(10C - 1.82C) = +3.68C	+21.68C

0 = +62C - 19.18C(P/F,i%,1) - 18.36C(P/F,i%,2)
-17.55C(P/F,i%,3) - 12.73C(P/F,i%,4) - 13.71C(P/F,i%,5)
-12.89C(P/F,i%,6) - 6.07C(P/F,i%,7) - 8.40C(P/F,i%,8)
- 7.14(P/F,i%,9) + 21.68(P/F,i%,10)

i = approximately 16.0%. This is the after-tax cost to the lessee of the money made available by the leasing management.

18-16 continued

(Although there are two reversals in sign in the series that shows the prospective after-tax differences in cash flow, this is one of the types of case discussed on pages 605-606 of the text where it is good enough for practical purposes to use conventional methods of analysis.)

18-17

If the 10% investment tax credit were available with ownership but not with leasing, the difference in cash flow at zero date would be changed by the amount of the credit, namely 10C. The equation for the solution to Problem 18-16 is changed to:

$0 = +52 - 19.18C(P/F,i\%,1)$ . . etc.

i = approximately 23.7%. That is, the availability of the tax credit with ownership and not with leasing has increased the after-tax cost of the funds made available by leasing from 16.0% to 23.7%.

Of course if the tax credit is available either with leasing or with ownership, the funds made available by leasing will be the same 62C calculated in the solution to Problem 18-16 and the after-tax cost of the money will continue to be 16.0%.

18-18

Where the terms of the lease are unaffected by the prospect of inflation, this prospect increases the attractiveness of leasing. If the figures in the final column of the table in the solution to Problem 18-16 are expressed in units of constant purchasing power, each figure from years 1 to 10 will be less than the one shown in the table. The computed after-tax cost of the money made available by leasing will therefore be less than the 16.0% calculated in the solution to Problem 18-16. The greater the expected rate of inflation, the greater the reduction in this computed cost of money where analysis is made in monetary units of constant purchasing power.

18-19

With a given set of terms of a lease, the prospect of inflation makes leasing less attractive to the lessor for the same reason that it becomes more attractive to the lessee. (See the answer to Problem 18-18.) As inflation continues, the manufacturer presumably will make his annual rental charges a larger fraction of the first cost of any machine. Leasing prices tend to rise during periods of inflation for the same reason that interest rates tend to rise during such periods.

18-20

(a) The following disbursements tend to increase the city's equity in the water system:

New fixed assets:	New well	$13,300
	Pump house and pump	9,600
	New water mains	12,800
Reduction of liability:	Repayment of bonds	25,000
		$60,700

The foregoing outlays may reasonably be viewed as "financial provision for depreciation." Their total is higher than the city engineer's estimate of $45,270 for the depreciation chargeable in 1981.

(b) The city engineer's estimate of the net present value of the plant on January 1, 1981 was $619,235. This is greater than the outstanding debt of $400,000. If the city engineer's estimate of net present value is accepted, the city had an equity of $219,235 in its water plant at the start of 1981. According to the problem statement this equity started at zero when the plant was constructed and has been built up entirely by revenues from the sale of water. Therefore, it appears that the overall financial provisions for depreciation up to the start of 1981 were more than adequate.

18-21

The cash flow differences between Plans P and R and their present worths at 12% and 15% now are:

	Cash flow difference	PW @ 12%	PW @ 15%
Date 0	-90N	-90.000N	-90.000N
Uniform flow 0-3	+24N per year	+61.032N	+58.800N
Date 3	+45N	+32.031N	+29.588N
Sum of present worths		+3.063N	-1.612N

$$i = 12\% + \frac{3.063N}{4.675N}(3\%) = 14.0\%$$

CHAPTER 19

Some Aspects of Economy Studies For Government Activities

General Notes:

The analysis of project proposals presented by different government agencies has become a subject of careful Congressional study and debate. The Joint Committee on Economy in Government (frequently called the "Proxmire Committee") has held numerous hearings and investigations of the current practices, and especially, the effects of the use of different interest rates by different agencies on the allocation of public funds. The desirability of using "opportunity costs" as the basis for setting the interest rates to be used in project justifications has been recognized by the committee, and so reported to Congress by the Committee. Some executive directives have been issued calling for the use of interest rates up to 10%, and considerable pressure is being put on all agencies to do a better job of applying sound economic principles to project proposal formulation and justification.

Therefore, this chapter is very important for students, whether they expect to work for government agencies or to be "just taxpayers." As interested citizens, their knowledge of proper analysis methods will enable them to influence government actions, and to demand better use of their tax dollars.

Good examples from local, state, and national project proposals can be found almost any time in the newspapers. Students can be assigned special projects to look into the methods and interest rates used in justifying the projects. They can attend public hearings and ask intelligent questions, and then prepare reports to present to the class. The reality of the principles presented in this chapter will stimulate a great deal of interest among the students and will help prepare them for an active role in government decision processes.

19-1

EUAC for government costs, Location X ($ x 1,000)

Initial cost	$6,776(.08)	= $542.08
Resurface	$160(11.3 mi)(A/F,8%,15)=$1,808(0.03683)	= 66.59
Maintenance	$4(11.3 mi)	= 45.20
	Total	$653.87

User costs, location X

fuel [(.15)($0.80)+(.85)($0.216)](700)(11.3)(365) = $876,539

time 700(.25)(11.3mi)(1/40mph)(60min/hr)($0.24)(365) = 259,844

Total $1,136,383

EUAC for government costs, location Y($ x 1,000)

Initial cost	$4,640(0.08)	= $371.20
Resurfacing	$160(13.5 mi)(0.03683)	= 79.55
Maintenance	$4(13.5 mi)	= 54.00
	Total	$504.75

User costs, location Y

fuel [(.15)($0.80)+(.85)($0.216)](700)(13.5)(365) = $1,047,192

time 700(.25)(13.5 mi)(1/40mph)(60min/hr)($0.24)(365) = $ 310,432

Total $1,357,624

Incremental B/C ratio X - Y:

$$B/C = \frac{\$1,357,624 - \$1,136,383}{\$653,870 - \$504,750} = \frac{\$221,241}{\$149,120} = \underline{1.48}$$

19-2

The interest rate at which the NAW of the cash flow differences equals 0 is the value of i at which the B/C ratio on the difference equals 1.0.

(Location X - Location Y)

$$\begin{aligned} 0 &= -\$2,136,000(i) + \$160,000(2.2\text{ mi})(A/F,i\%,15) \\ &\quad + \$4,000(2.2\text{ mi}) + (\$1,357,624 - \$1,136,383) \\ &= -\$2,136,000(i) + \$352,000(A/F,i\%,15) + \$221,241 \end{aligned}$$

$$\underline{i = 10.8\%}$$

19-3

(a) Overall B/C = $99,104/$46,853 = 2.12

Incremental B/C ratios for:

(b) Flood Control = $10,975/$8,110 = 1.35

(c) Irrigation = $54,875/$9,434 = 5.82

(d) Power = $12,622/$9,992 = 1.26

(e) Water = $16,462/$1,929 = 8.53

19-4

B/C for single purpose projects:

(a) Flood Control: $10,975/$20,000 = 0.55
(b) Irrigation: $54,875/$28,800 = 1.91
(c) Water: $16,462/$15,600 = 1.06

19-5

Capitalized cost of Plan I (1/3 capacity now, 1/3 at 10 years, and 1/3 at 35 years):

$2,000,000+$11,000/0.09+[$2,000,000+$11,000/0.09]
(P/F,9%,10)+[$2,000,000+$11,000/0.09](P/F,9%,35)
= [$2,000,000+$11,000/0.09][1+0.4224+0.0490] = $3,122,638.

Plan II (1/2 capacity now, 1/2 at 20 years):
[$2,400,000+$10,000/0.09][1+(P/F,9%,20)]
= $2,511,111(1.1784) = $2,959,093.

Plan III (2/3 capacity now, 1/3 at 35 years):
$2,700,000+$8,000/0.09+[$2,000,000+$11,000/0.09]
(P/F,9%,35) = $2,788,889 + $2,122,222(0.0490)
= $2,892,878.

Plan IV (full capacity now) = $3,400,000.

Plan III is the most economical.

19-6

(a) B/C = ($800-$300)/$300+$100) = 1.25.
(b) B/C = $800/($300+$100+$300) = 1.14.
(c) B/C = ($800-$300-$100)/$300 = 1.33.
(d) B-C = $800-$300-$300-$100 = $100.

19-7

For LD compared to NFC; ($000)omitted:

B/C = ($410-$155)/[$2,000(0.06344) + $28] = 1.65

For HD compared to LD; ($000) omitted:

B = $155-$20-$15 = $120
C = $1,300(0.06344) +$19 = $101.5; B/C = 1.18.

Thus the high dam should be selected since the marginal benefits are sufficient to cover the marginal costs. The overall B/C ratio for HD is:

B/C = ($410-$20-$15)/[$3,300(0.06344)+$47] = 1.46.

19-8

For LD compared with NFC; ($000 omitted):

$$B/C = (\$410-\$155)/[\$2,000(0.10086)+\$28] = \underline{1.11}$$

For HD compared with LD; ($000 omitted):

$$B/C = (\$155-\$20-\$15)/[\$1,300(0.10086)+\$19] = \underline{0.80}.$$

Thus LD is preferred to HD since the marginal benefit is not sufficient to cover the marginal cost. The overall B/C for HD is:

$$B/C = (\$410-\$20-\$15)/[\$3,300(0.10086)+\$47] = \underline{0.99}.$$

19-9

For LD compared with NFC; ($000 omitted)

$$C = \$2,000(0.06344)-\$1,500(0.00344)+\$28 = \$149.7$$

$$B/C = (\$410-\$155)/\$149.7 = \underline{1.70}.$$

For HD compared with LD; ($000 omitted):

$$C = \$1,300(0.06344)-\$975(0.00344)+\$19 = \$98.1$$

$$B/C = (\$155-\$20-\$15)/\$98.1 = \underline{1.22}.$$

Thus HD is preferred to LD since the marginal benefit exceeds the marginal cost. The overall B/C for HD is:

$$B/C = (\$410-\$20-\$15)/[\$3,300(0.06344) - \$2,475(0.00344) + \$47] = \underline{1.51}.$$

19-10

For A compared to NFC; ($000 omitted):

$$C = \$12,000(A/P,8\%,50)-\$1,000(A/F,8\%,50) + \$30$$
$$= \$12,000(0.08174)-\$1,000(0.00174)+\$30 = \$1,009$$
$$B/C = (\$1,700-\$900+\$40)/\$1,009 = \underline{0.83}.$$

For B compared to NFC; ($000 omitted):

$$C = \$15,000(0.08174)-\$2,500(0.00174)+\$40 = \$1,262$$
$$B/C = (\$1,700-\$600+\$60)/\$1,262 = \underline{0.92}.$$

For C compared to NFC; ($000 omitted):

$$C = \$18,000(0.08174)-\$3,500(0.00174)+\$50 = \$1,515$$
$$B/C = (\$1,700-\$300+\$120)/\$1,515 = \underline{1.003}.$$

The only project for which benefits exceed costs is Dam C.

19-11

Henry Steiner's thesis (Report EEP-17,P.32) gives these figures as:

1.	Palmillas-Miquihana (3)	1602 pesos per inhabitant served
2.	Las Norias-Cruillas (1)	1761 " " " "
3.	Mendez-Entronque (4)	1,987 " " " "
4.	El Capulin-Bustamente (2)	2,068 " " " "

Because this is an inverse cost effectiveness index, the lowest figure is the best and the priority list starts with Palmillas-Miquihana.

19-12

Steiner's EEP-17, p.101, gives these figures as:

1.	Casas-Soto La Marina-La Pesca	(4)	2.11
2.	Llera-Gonzalez	(6)	2.02
3.	El Barretal-Padilla	(8)	2.01
4.	Mendez-Burgos	(10)	1.99
5.	El Barretal-Sta. Engracia	(3)	1.82
6.	Altamira-Aldama	(2)	1.58
7.	Jimenez-Abasdo	(9)	1.50
8.	Ebano-Manuel	(7)	1.21
9.	Limon-Ocampo	(5)	1.15
10.	Hidalgo-La Mesa	(1)	0.90

19-13

Daily water use is = 40,000(150) = 6 mgd

Plant capacity = 2(6mgd) = 12 mgd
Plant investment = (12mgd)($35,000) = $420,000

Annual bond redemption = $420,000/20	=	$21,000/yr
First year interest = $420,000(0.07)	=	29,400
Annual maintenance = 0.03($420,000)	=	12,600
Annual chemicals = (320-70ppm)(6mgd)($0.15)(365)	=	82,125
Annual pumping = $2.40(6mgd)(365)	=	5,256
Annual labor	=	12,000
Total first-yr cost		$162,381

Increased cost/1,000 gal =
$162,381/(6,000)(365) = $0.0741/1,000 gal

Annual savings to citizens:

soap; 40,000(38.5-30.8)($0.45)	=	$138,600
chemicals; $0.29(110)(320-70)	=	7,975
heater replacement; $140(4,000)[(A/P,6%,8)-(A/P,6%,16)*]	=	34,770
Total		$181,345

19-13 continued

*interest rate of 6% for the citizen is arbitrarily chosen as a reasonable one for personal projects.

The annual saving of $181,345 is more than the first year's cost. Since the annual cost will decrease by $1,470/year throughout the 20 years of the bonds, the project is desirable from the standpoint of the public interest.

19-14

Here is a problem in which the examiner deliberately omitted information that he thought was relevant. A guess at the solution expected by the examiner is as follows:

Any decision between (1) spend extra money now, and (2) spend extra money in 10 years calls for an estimate of the money to be spent in 10 years as well as the amount to be spent now. The examiner gave only (1) and presumably expected the examinee to estimate (2) in part from common sense and observation. Clearly, different types of estimates are appropriate for each of the three stated elements of first cost. In each analysis, it is desirable to have in mind that (F/P,4%,10) = 1.4802, or about 1.5. Unless the extra expenditure now to provide for six lanes seems likely to prevent an expenditure at least 1.5 times as great in 10 years, it will not be economical to make the extra present expenditure.

With regard to right of way, it is a common observation that commercial developments occur along a new highway. When a highway is widened, a major element of cost is for right of way. It seems reasonable that the extra expenditure of $80,000 now to provide right of way for 6 lanes will prevent an expenditure of many times that amount in 10 years.

Moreover, it seems reasonable that the extra $240,000 necessary to do grading and build drainage structures for 6 lanes at once will prevent expenditures of much more than $360,000 in 10 years. (These are presumably the main items in "all other elements of cost.") The widening of bridges and culverts tends to be relatively expensive compared to their initial cost, and the economics of scale in doing the entire grading job at once are quite substantial. It is significant that provision for 50% more capacity in "all other elements of first cost" adds only 20% (i.e., $240,000/$1,200,000) to the first cost of these elements.

On the other hand, there are evidently no economics of scale with respect to pavement construction because an increase from 4 to 6 lanes will increase the first cost by 50%. In addition, 2 extra lanes of pavement built now will presumably deteriorate during the 10 years that will elapse before they are needed and therefore will be less desirable than 2 new lanes built 10 years hence. (There will be no such deterioration of right of way purchased now and presumably negligible deterioration of grading.)

9-14 continued

The answer to the question is therefore that $320,000 in addition to the cost of the 4-lane highway can be spent economically at once and that this should be divided $80,000 to right of way and $240,000 to elements of first cost rather than pavement.

The foregoing examination question was one used in the late 30's. A modern view of public works economy is that prospective "consequences to whomsoever they may accrue" are relevant. In this case of the new highway, highway users would presumbably be unaffected by the decision to buy right of way at once and to do the grading and build drainage structures sufficient for 6 lanes. Therefore, the foregoing analysis would not be changed by the broader viewpoint. However, a consideration of the merits of a proposal to pave all 6 lanes at once would require estimates and evaluations of the expected consequences.

19-15

The "remaining justifiable expenditure" is the smaller of either benefits or alternative costs less the separable costs. The allocation of nonseparable costs is based on the percentage of the remaining justifiable expenditure for the individual purpose is of the total remaining justifiable expenditure. Therefore, it is not necessarily true that the purpose having the larger incremental B/C will always be allocated a larger percentage of the nonseparable costs. Under usual situations, however, the expectation is that a larger portion of the nonseparable costs will be allocated to the purpose with the larger incremental B/C, simply because the larger the benefits, the larger the alternative cost is likely to be.

The rational foundation for this method stems from the fact that most multipurpose projects grow out of an original single purpose project proposal. Commonly it is found that by adding some additional investment to the original project, additional purposes can be served. A certain capacity dam might well serve for flood control purposes, but to add power generation, and irrigation, additional expenditures for increased height and strength, more land to be inundated, flumes and channels, etc. must be made. These adtional expenditures cannot be entirely separated to the various purposes. The proposed method, in effect, says, "separate everything you can and allocate the remaining on the basis of the separate benefits to be enjoyed by the individual purposes." Basically this is the concept of basing charges on ability to pay.

19-16

(See table of Problem 19-17)

The capitalized costs and capitalized benefits based on 7% are shown in that table. Using values from that table, the B/C ratios are:

Total	;	\$64,425/\$44,280 = 1.45
Flood Control	;	\$ 7,135/\$ 8,114 = 0.88
Irrigation	;	\$28,800/\$ 8,812 = 3.27
Power	;	\$ 8,205/\$ 8,883 = 0.92
Water	;	\$10,702/\$ 2,014 = 5.31

The B/C ratios for Flood Control and Power are less than 1.0, thus these purposes will not pay for themselves.

19-17

The table is shown on the following page. The operation, maintenance, and replacement costs and the benefits change substantially with a change in interest rate. Those purposes that have low annual costs but high construction costs have a greater reduction in the B/C ratio than those with low construction costs. In this problem, an increase in interest rate caused a decrease in all B/C ratios with two going below 1.0; Flood Control and Power. Consequently, these purposes are suspect and may be candidates to be eliminated.

The actual B/C ratios are given in Problem 19-16. Note that, in the allocation of joint costs, no allocation is made to those purposes which are unable to pay for themselves.

19-18

(a) Flood Control

If no nonseparable costs were allocated to flood control, the i at which B/C = 1 can be found by:

$$\$500(P/A,i\%,100) = \$8,000 + \$110(P/A,i\%,100)/(P/A,4.5\%,100)$$
$$\$500(P/A,i\%,100) = \$8,000 + (\$110/21.950)(P/A,i\%,100)$$
$$(P/A,i\%,100) = \$8,000/\$494.99 = 16.162$$
$$\underline{i = 6.2\%}$$

(b) With same assumption, let i = the break even interest rate for power:

$$\$575(P/A,i\%,100) = \$6,700+\$3,292(P/A,i\%,100)/(P/A,4.5\%,100)$$
$$(P/A,i\%,100) = \$6,700/\$425.02 = 15.7639$$
$$\underline{i = 6.36\%}$$

<u>19-17</u> (P/A,7%,100) = 14.269 ; Figures in $ x 1,000

Item*	Flood Control	Irrigation	Power	Water	Fisheries	Navigation	Total
1							44,280
a							(39,500)
b							(4,780)
2	7,135	35,672	8,205	10,702	1,284	1,427	64,425
3	20,000	28,800	12,622	15,600	-	-	-
4	7,135	28,800	8,205	10,702	1,284	1,427	57,553
5	8,114	8,812	8,883	2,014	-	-	27,823
a	(8,000)	(6,800)	(6,700)	(1,600)	-	-	(23,100)
b	(114)	(2,012)	(2,183)	(414)	-	-	(4,723)
6	[- 979]	19,988	[- 678]	8,688	1,284	1,427	31,387
7	-	63.7%	-	27.7%	4.1%	4.5%	100%
8	-	10,483	-	4,559	674	741	16,457
a	-	(10,430)	-	(4,576)	(656)	(738)	(16,400)
b	-	(36)	-	(16)	(2)	(3)	(57)
9	8,114	19,295	8,883	6,573	674	741	44,280
a	(8,000)	(17,230)	(6,700)	(6,176)	(656)	(738)	(39,500)
b	(114)	(2,048)	(2,183)	(430)	(2)	(3)	(4,780)
10	8	141	153	29	2	2	335

*Items are same as in Table 19-1

19-19

(a) 4.5%, perpetual life:
B = \$575/0.045=\$12,778; C=\$1,600+\$329/(P/A,4.5%,100)(0.045)
= \$1,600 + \$242 = \$1,842
B/C = \$12,778/\$1,842 = 6.94.

(b) 4.5%, 50 year life:
B = \$575(P/A,4.5%,50) = \$11,363
C = \$1,600+\$329(P/A,4.5%,50)/(P/A,4.5%,100)
= \$1,600 + \$296 = \$1,896
B/C = \$11,363/\$1,896 = 5.99.

(c) 4%, 100 year life:
B = \$575(P/A,4%,100) = \$14,090
C = \$1,600+\$329(P/A,4%,100)/(P/A,4.5%,100) = \$1,967
B/C = \$14,090/\$1,967 = 7.16.

(d) 5%, 100 year life:
B = \$575(P/A,5%,100) = \$11,413
C = \$1,600+\$329(P/A,5%,100)/(P/A,4.5%,100) = \$1,897
B/C = \$11,413/\$1,897 = 6.02.

19-20

All monetary figures in Table 19-1 are present worths. If the uniform annual worth figures had been used, all the monetary figures would have been multiplied by (A/P,4.5%,100) = 0.04556. No change would have occurred in the B/C ratios.

19-21

This is another possible problem for class discussion. A similar topic was used as an illustration in the introductory material regarding "benefits" near the beginning of Chapter 9.

19-22

	Average daily traffic costs:	
	4-Way Stop	Signals
Operating	\$41.60	\$29.10
Time delay	33.40	19.40
Accident	6.80	10.30
	\$81.80	\$58.80

Annual benefit from signalization = 365(\$81.80-\$58.80) = \$8,395
EUAC of signalization = \$29,500(A/P,10%,12)+\$1,600 = \$5,929

B - C = \$2,466; B/C = 1.42.

19-23

	4-Way Stop	2-Way Stop	Signal
ADT user costs	$81.80	$66.80	$58.80
Annual user costs	$29,857	$24,382	$21,462
EUAC, gov't @10%	0	0	$ 5,929

The change to the two-way stop produces annual benefits of $5,475 at no government cost. Actually, in Problem 19-22, all alternatives were not formulated and evaluated. (Another alternative would be to remove the stop signs altogether.) Thus the signalization should be compared to the 2-way stop.

B/C = ($24,382 - $21,462)/$5,929 = 0.49

The signalization is not economic as compared to the 2-way stop.

19-24

Consideration of: (1) noncommercial time and (2) atmospheric pollution would increase the benefits derived both by the 2-way stop and signalization. Thus the change to the 2-way stop would certainly be enhanced. Evaluation of signalization vs. 2-way stop would require an increase in annual benefits of $3,009, or 103%, to produce a B/C ratio of 1.0. This might occur. However, this particular signalization would still have to compete with many similar projects, evaluated on the same basis, for limited public funds.

19-25

This is a possible class discussion problem to illustrate one type of conceptual difficulty that sometimes arises in the economic analysis of proposed public works. In the view of the authors, the $1.24 figure is the correct one to use rather than the $1.35 figure.

If a public investment, such as a highway improvement, has the incidental effect of reducing the taxes paid by a portion of the public, such tax reduction should not be regarded as a project "benefit." Presumably, taxes are levied to meet the needs of governments, and the total amount of the taxes levied will be based on such needs as they appear to the people and their elected representatives. In the long run, a tax reduction in one place is likely to be made up by additional taxes collected in some other place.

19-26

174 ft. vs. no dam:

B = 19,450($37.50) = $729,375
C = $5,865,700(A/P,9%,40) + $124,000 = $669,275
B/C = $729,375/$669,275 = 1.09.

194 ft. vs. 174 ft.:

B = 21,675($37.50) - $729,375 = $83,437.5
C = $6,467,700(0.09296)+$142,000-$729,375 = $13,862
B/C = $83,437.5/$13,862 = 6.02.

211 ft. vs. 194 ft.

B = 23,600($37.50) - $812,812.5 = $72,187.5
C = $7,354,400(0.09296)+$169,000-$743,237 = $109,428
B/C = $72,187.5/$109,428 = 0.66.

The 194 ft. dam is the most economic. Its overall B/C ratio is $812,812.5/$743,237 = 1.09 and it produces electric power for $743,237/21,675 = $34.29/kw-year.

CHAPTER 20

Some Aspects of Economy Studies For Regulated Businesses

General Notes

Regulation has become so all pervasive that no business, regardless of how small, can escape the necessity to comply with some kinds of government regulations. Many regulations can be met by anyone of several alternative methods with the result that a business firm should analyze the life-cycle costs of each alternative method before making a decision as to the method to be adopted. There are many examples of such decision opportunities in the engineering journals, THE WALL STREET JOURNAL, and other publications which the teacher can use to increase the interest of his or her class in the use of engineering economy as applied to regulatory matters. One case study with several problems is included in the problems for this chapter.

Engineering economy problems relating to regulated public utilities, whether privately or publicly owned, are rather special and require a thorough knowledge of the rules and policies specified by the regulatory body. It is suggested that the teacher obtain information from the local regulatory agency in order to make the examples more appealing to the students. Does the agency use the concept of minimization of revenue requirements, is it liberal or very tight regarding rate base and fair rate of return, does it insist on flow-through of investment tax credits and benefits of liberalized depreciation? All of these points can be used to help explain the application of engineering economy principles to the special problems of regulated industries. Since all persons are either willing or unwilling users of public utilities, the students usually become interested in how their utility companies make investment decisions and how they affect the rates they have to pay for the service. This provides the opportunity to assign groups of students to study rate cases and investment proposals for independent project reports.

Giving the students some information regarding the annual capital investments by utility companies in your area, and data on income and operating expenditures as well as cases and hearings before the regulatory bodies will stimulate interest. Since the utilities are so capital intensive, there are many opportunities for the application of the principles of engineering economy to such widely different matters as proposed capital investments by the utility companies in plant expansion or whether or not the stock of the local utility company is an attractive investment.

The authors understand that this chapter is not ordinarily used in a first course in engineering economy and more often used in a second or advanced course. We would suggest that the topic of regulation (in the general sense) be at least introduced in a first course because many of the students who will not take the second course will find themselves dealing with the economics of regulation very soon after graduation. Perhaps just the use of Example 20-1 and a discussion of the engineering economy aspects of the case will be enough to make the students aware of the kind of problems they may soon face.

20-1   Plan I

For land: n= 40; s = 100%; P = $10,000
t = [0.52/(1-0.52)](0.10-0.08(0.40))[1+(1-1)(1-(A/G,10%,40)/40)]
  = (0.52/0.48)(0.10-0.32)(1) = 0.073667

For building and conduit: n= 40; s= 0; P= $200,000
t = (0.52/0.48)(0.10-0.80(0.40))[0+(1-0)(1-(A/G,10%,40)/4)]
  = 1.083333(0.0680)(1- 9.096/40) = 0.05691

For cable: n= 20; s = 20%; P = $40,000
t = 1.083333(0.0680)[0.20 +(1-0.20)(1 - (A/G,10%,20)/20)]
  = 1.083333(0.0680)(0.20 + 0.80(1 - 6.508/20) = 0.054490

First installation:		
CR: $10,000(0.10)=	$1,000	
200,000(A/P,10%,40)=	20,452	
32,000(A/P,10%,20) + $8,000(0.10)=	4,559	
EUACF for income tax:		
$10,000(0.073667)=	737	
200,000(0.056915)=	11,383	
40,000(0.054490)=	2,180	
Ad valorem tax: $250,000(0.03)=	7,500	
O & M Disbursements=	30,000	
Annual revenue required	$77,811	
For second installment: n=20; s= 20%; P =$60,000; t = 0.054490		
CR:		
$48,000(0.11746)+ $12,000(0.10) =	$6,838	
EUACF for income tax:		
$60,000(0.054490)=	3,269	
Ad valorem tax: $60,000(0.03)=	1,800	
O & M disbursement =	8,000	$19,907

20-1 Continued:

The third installation is the same as the second except that it occurs in 14 years.

Revenue required for first 20 years:

= \$77,811 +\$19,907(P/A,10%,12)(P/F,10%,8)(A/P,10%,20)
+19,907(P/A,10%,6)(P/F,10%,14)(A/P,10%,20)
= \$77,811 +\$19,907(6.814)(0.4665)(0.11746)
+\$19,907(4,355)(0.2633)(0.11746) = \$87,925

Plan II

t's for land, building and conduit, and cable are the same.

CR: \$10,000(0.10)=	\$ 1,000
200,000(0.10226)=	20,452
80,000(0.11746) +\$20,000(0.10)=	11,397
UACF for income tax:	
\$10,000(0.073667)=	737
200,000(0.056910)=	11,382
100,000(0.054490)=	5,449
Ad valorem tax:\$310,000(0.03)=	9,300
O & M disbursements=	38,000
Total revenue required	\$97,717

20-2

(a) Revenue required if leased: i = 8% (Since payments are monthly, use continuous compounding factors for effective 8%.)

Year		
0-1	\$2,400 x 0.9625=	\$2,310
1-2	2,400 x 0.8912=	2,139
2-3	1,680 x 0.8252=	1,386
3-4	1,680 x 0.7641=	1,284
		\$7,119

\$7,119(A/P,8%,4) = \$2,149

(b) Revenue required if purchased:

t = (0.52/0.48)(0.08 - 0.06(0.50)) [0.10 +(1-0.10)
(1 - (A/G,8%,4)/4)] = 0.03686

Total revenue required annually = \$9,000(A/P,8%,4) +
\$1,000(0.08) +\$10,000(0.03686) = \$3,166

(c) Revenue required if leased and in rate base:

t = (0.52/0.48)(0.08 - 0.06(1.00))[0.10 + (1-0.10)
(1 - (A/G,8%,4)/4) = 0.01474

Total revenue required annually = \$9,000(A/P,8%,4) +
\$1,000(0.08) + \$10,000(0.01474)] = \$2,944

Comments: In the first case, the consumer receives the service without paying the company anything other than the actual cost of

20-2 Continued:

the service and without paying any income tax.

If the truck is included in the rate base the consumer is paying 8% on the beginning of year value, based on straight line depreciation and the income tax incurred by the company. With 100% financing, (c), the income tax is quite small, however the income tax increases if the truck is only 50% financed by debt.

It turns out that by leasing without having the truck in the rate base, the consumer is paying less than straight line depreciation ($9,000/4 = $2,250.) This means that the leasing company must have a "gimmick." Its purchase price may be less than for the utility company and/or its resale price may be higher than the utility company estimates.

Leasing without the item being included in the rate base seems unfair to the utility company because it is not compensated in anyway for taking the risk involved in the acquisition of the truck under a fixed lease-price contract.

20-3 (a) Embedded cost of debt, June 1980, before refunding maturing bond = $i = (20/75)(4.4) + (20/75)(4.9) + (20/75)(5.8) + (15/75)(8.5) = \underline{5.73\%}$

(b) Embedded cost of debt after selling new bonds and refunding the maturing bond = $i = (20/85)(4.9) + (20/85)(5.8) + (15/85)(8.5) + (30/85)(9.8) = \underline{7.48\%}$

20-4 (a) Cost of capital = $7.48\%(0.55) + 12.5\%(0.45) = \underline{9.74\%}$

(b) Current cost of capital = $9.8(0.55) + 12.5(0.44) = \underline{11.02\%}$

(c) If the company is allowed to use the current cost of capital as the fair rate of return it would be able to raise its service price rates.

If the fair rate of return is based on the embedded cost of capital, the company's stock will probably decline on the market because the fair rate of return will be less than the current cost of debt.

20-5 Debt/Equity ratio = 55/45:

For land: $n = 40$; $s = 100\%$, $P = \$10{,}000$ (See solution to 20-1)

$t = 1.083333(0.10 - 0.08(0.55))(1) = \underline{0.06067}$

For building and conduit: $n = 40$, $s = 0$; $P = \$200{,}000$

$t = 1.083333(0.0560)(0 + (1-0)(1 - 9.096/4)) = \underline{0.04687}$

For cable: $n = 20$; $s = 20\%$; $P = \$100{,}000$

$t = 1.083333(0.0560)(0.20 + (1 - 0.20)(1 - 6.508/4)) = \underline{0.044874}$

Only the revenue requirements for income tax change:

20-5 Continued:

	55/45	40/60
\$ 10,000(0.06067)=	\$ 606.70	\$ 737
200,000(0.04687)=	9,374.00	11,382
100,000(0.04487)=	4,487.40	5,449
	\$14,468.10	\$17,568

(b) Interest on debt = 6%; fair rate of return = 9%

For land:
t = (0.52/0.48)(0.09 - 0.06(0.40))(1) = 0.07150

For building and conduit:
t = (1.083333)(0.0660)(0 + (1-0)(1 - 9.796/40)) = 0.053990

For cable:
t = (1.083333)(0.0660)(0.20 +(1-0.20)(1 - 6.767/20)) = 0.052146

Revenue required:

CR: \$10,000(0.09)=	\$ 900
200(A/P,9%,40) =	18,592
80,000(A/P,9%,20) + \$20,000(0.09)=	10,564
Ad valorem tax: \$310,000(0.03) =	9,300
O & M disbursements=	38,000
Income tax:	
\$10,000(0.07150)=	715
200,000(0.053990)=	10,798
100,000(0.052146)=	5,246
Total revenue required annually=	\$94,115

(c) Increasing the debt/equity ratio with the same cost of debt and the same fair rate of return reduces the revenue required becuase the company pays out more in interest (a tax deductible expense) and has less taxable income on which to pay income tax.

Decreasing the cost of debt and the fair rate of return means the company must lower its service rates but also pays less interest. The effect is to reduce the revenue requirement.

20-6 No change is necessary for the first installation since it will serve for 20 years. A new t must be computed for the second installation, based on the new "salvage value."

t = (0.54/0.46)(0.10-0.09(0.50))[0.595 +(1-0.595)
(1 - (A/G,10%,12)/12)] = 0.055003

Revenue required for second installation:

CR: (\$160,000-\$95,138)(A/P,10%,12) + \$95,138(0.10) =	\$19,033
Ad valorem tax: 160,000(0.02)=	3,200
O & M disbursements:	8,000
Income tax: \$160,000(0.055003)=	8,800
	\$39,033

20-6 Continued:

Equivalent annual revenue required = \$34,447 +\$39,033(P/A,10%,12)
(P/F,10%,8)(A/P,10%,20) = \$34,447 + \$14,574
= \$49,021

This method of dealing with the cotermination problem divides the second installation into two parts, the first part, 12 years, to be used in the study period of 20 years.The residual (or "salvage") value is computed as the present worth at time 20 of the unrecovered investment. The use of that residual value results in a slight increase in the revenue requirements in the 20 years.

20-7 A new t must be computed based on the estimated salvage value for the second installation :

t = (0.54/0.46)(0.10 - 0.90(0.50) 0.52 +(1-0.52)
(1 - (A/G,10%,12)/12) = 0.050161

Revenue required:

CR: (\$160,000-\$83,200)(A/P,10%,12) +\$83,200(0.10) =	\$19,591
Ad valorem tax: \$160,000(0.02)=	3,200
O & M disbursements:	8,000
Income tax: \$160,000(0.050161)=	8,026
	\$38,817

Equivalent annual revenue required = \$34,447 +
\$38,817(P/A,10%,12)(P/F,10%,8)(A/P,10%,20) = \$48,941

Revenue required :	Example 20-2	\$48,292
	Problem 20-6	49,021
	Problem 20-7	48,941

It can be seen that the effort to allocate the costs of the first 12 years of service of an asset with an estimated life of 20 years is slightly sensitive to the method used to estimate the residual value at the end of the study period. The second two methods will result in higher revenue requirements in the first period, but the revenue requirements for the last 8 years of the asset's life will have a lower revenue requirement. Some utility companies prefer the second method.

Problems 20-8 through 20-15 : NOTE WELL : The solutions to these problems were based on the income tax laws in force before the Economic Recovery Tax Act of 1981.

The case study did not specify anything about the depreciation methods to be used or the i*. The assumptions made are as follows:
Straight line depreciation will be used for building, equipment, and chemical treatment plant. 10-year life is assumed for equipment, 20-year for building and chemical plant.

Investment Tax Credit (ITC) of 10% applies to equipment and chemical treatment plant.

Gross sales (income) = \$600,000 and taxable income = \$75,000. Thus, \$600,000 - \$75,000 -(depreciation, salaries and wages, and property taxes) = A,O & M disbursements. (A,O & M = administrative, marketing, supplies, materials, operating and maintenance disbursements.) Assume A,O & M disbursements will not change except for addition of \$5,000 for chemical treatment plant if the existing plant is retrofitted. A,O & M will decrease by \$15,000 minus \$5,000 for chemical treatment plant if the corporation moves to the other state.

The corporation has adequate working capital to meet any required investments over the \$200,000 loan for retrofitted plant.

Preliminary calculations for retrofitted plant :

Compute A,O & M disbursements:

Depreciation-building= \$250,000/30=	\$8,333
Depreciation- equipment= \$125,000/12=	10,417
President's salary	40,000
Unskilled labor= 10x\$8x2,000hrs./yr.=	160,000
Skilled labor= 4x\$15x2,000=	120,000
Property taxes= \$300,000(0.015)=	4,500
Total	\$343,250

A,O &M disbursements = \$600,000 -\$75,000-\$343,250 = \$181,750

Current value of existing plant and the net realizable cash flow if sold now :

		Book Value	Gain (Loss)
Land	\$100,000	\$50,000	\$50,000
Bldg.	50,000	125,000	(75,000)
Equipment	15,000	15,000	0

Since a corporation can only use long term capital losses to balance long term capital gains or short term capital gains, the tax advantage of the net loss of \$25,000 cannot be realized if plant is sold, because there will be no long term capital gains until after the carry-forward time limit expires.

Capital required for rehabilitation of existing building:

Building rehab	\$190,000
Chem. Treat. plant	50,000
New equipment	140,000
Total	\$380,000
Less ITC	19,000
Net required	\$361,000

Sources of capital:

Disposal of old equipment	$15,000
Equipment fund	125,000
Loan from bank	200,000
From working capital	21,000
Total	$361,000

A major question is: Can the ITC be realized immediately?

Tax liability for current year:

State= 0.05($50,000) +0.08($25,000) = $4,500

Federal taxable income = $75,000 - $4,500 = $70,500

Federal tax = 0.17($25,000) +0.20($25,000) + 0.30($25,000) = $15,400

Thus, only $15,400 can be used this year (assuming same taxable income) leaving $3,600 to be carried forward to next year. This is obviously an approximation but it is close enough for our purposes.

Depreciation charges for rehabilitated plant:

Building:($125,000 + $190,000)/20 =	$15,750
Equipment: $125,000/10 =	12,500
Chem. treat. Plant: $50,000/20=	2,500
Total	$30,750

Terminal salvage value of plant and equipment 20 years hence:

Land	$140,000	($90,000 long term gain)
Equipment	15,000	
Total	$155,000	

Cash Flow before debt service and income taxes:

Income		$600,000
Less:		
Salaries and wages	$320,000	
Property taxes	4,500	
A, O & M	181,750	
Chem. plant operation	5,000	511,250
Net cash flow		$88,750

Headings for columns of cash flow tables in 20-8 and 20-13:

A Cash flow before debt service and income taxes

B Cash flow for debt repayment

C Cash flow for interest payments

D Depreciation

E Taxable income = A + C + D

F Cash flow for income taxes = D x appropriate tax rates.

G Cash flow after income taxes and debt service= A + B + C + F

20-8

Yr.	A	B	C	D	E	F	G
0	-$165,000 (existing plant value)					)	
0	- 365,000 (new investments)					) +$15,400	-$314,600
0	+ 200,000 (bank loan)					)	
1	+ 88,750	-$40,000	-$24,000	-$30,750	+$34,000	- 3,810*	+ 20,940
2	+ "	- "	- 19,200	- "	+ 38,800	- 8,562	+ 20,988
3	+ "	- "	- 14,400	- "	+ 43,600	- 9,754	+ 24,596
4	+ "	- "	- 9,600	- "	+ 48,400	- 10,866	+ 28,284
5	+ "	- "	- 4,800	- "	+53,200	- 12,139	+ 31,811
6-10	+ "	0	0	- "	+58,000	- 13,848	+ 74,902
10	- 125,000(new equipment)					+ 14,000	- 111,000
11-20	+ 88,750			- 30,750	+58,000	- 13,848	+ 74,902
20	+ 155,000					- 10,800	+ 144,200

* State + Federal Taxes - ITC = $1,700 + $5,710 -$3,600 carried forward = $3,810
Capital gains tax in yr. 20 = 0.30(0.40)($90,000) = $10,800

Tax rates:

State:		Federal:	
First $50,000	@ 5%	First $25,000	@17%
Over $50,000	@ 8%	Next $25,000	@20%
		Next $25,000	@30%
		Next $25,000	@40%
		Over $100,000	@46%

20-9

Compute PW of After-Tax Cash Flow at 13% and at 14%: PW @ 13% = +$14,560
PW @ 14% = - $10,720. By interpolation, rate of return on corporation's equity is 13.6%, which should be compared with the Corporation's i*.

20-10

Examination of the cash flow table in Problem 20-8 shows that there will be a net positive cash flow each year except for the tenth year when the equipment must be replaced. Yes, the corporation can meet the cash flow requirements for debt repayment and payment of interest if the gross income remains the same. The only crunch may come at time zero if the working capital reserve is not great enough to provide about $25,000 for capital investment.

20-11

Current realizable value of existing plant is $100,000 + $50,000 + $15,000 = $165,000 (for land, building and old equipment.) Available working capital is not affected by the sale of the old plant.

20-12

Rehabilitation of the existing plant plus the chemical treatment plant in order to meet regulatory requirements adds ($190,000+ $50,000) = $240,000 to capital investment to be recovered in 20 years and $5,000 per year in operating disbursements.

EUAC of compliance = $240,000(A/P,20%,20) +$5,000 =
$49,286 + $5,000 = $54,286 per year.

A factor not considered in a before-tax analysis is that the existing building was being depreciated on the basis of a 30-year remaining life from time of purchase. This increases the remaining life from 15 years to 20 years, thereby reducing the annual depreciation charge on the original investment.

20-13

Preliminary calculations for proposal to move plant:

Investment required:

Land	$50,000
Building	250,000
Chem.Treat. Plant	50,000
Equipment	40,000
Total	$490,000

One-time expenses for move (all deductible expenses):

Moving office, tools and dies	$40,000
Moving skilled employees	20,000
Extra training=10($6)2,000/12=	10,000
Total	$70,000

Funds needed immediately= $541,000

Source of funds:

Sale of plant	$150,000	New Equip. fund	$125,000
Sale of equip.	15,000	Loan	250,000
Working capital	1,000	Total	$541,000

Net cash flow before debt service and income taxes:

Gross sales			$600,000
Less:			
Salary	$40,000		
Skilled	132,000		
Unskilled	120,000		
A, O& M	171,750		463,750
	Net cash flow yr.1-5		$136,250
Less moving expenses, time zero			70,000
Cash flow, yr. 1			$66,250
Property taxes, year 6 through 40= ($300,000 +$50,000 +$ 140,000)(0.015)=			$7,357
Depreciation:			
Building : $250,000/40=		$6,250	
Chem treat. plant:$50,000/20 =		2,500	
Equipment: $125,000/10 =		12,500	
Total depreciation		$21,250	

Cash Flow Table on next page.

20-14

Property tax foregone =($300,000 +$140,000)(0.015)= $6,600/yr.

Sales tax increase= (0.05)(0.4)(salaries and wages)=
= (0.05)(0.4)($292,000) = %5,840/yr.
Assume 40 year life of sales tax increase.
PW = 0 = -$6,600(P/A,i,5) +$5,840(P/A,i,40)
try 50%: PW = -$11,462 + $ 11,680 = +$218
i is very close to 50%

The state is making a very good deal in that it will "earn" over 50% on its investment in the property tax incentive.

The state will also receive a substantial increase in state income tax from the increased salaries and wages earned. The business firms in the area will also enjoy increased sales and additional profits, adding, inturn, to the income tax they will pay to the state. On the other hand, some of the costs of service to citizens (schools, police, fire protection, etc.) will be increased, butthe state will also probably have lower welfare payments to its present citizens.

20-15

i = 14% after income tax.

EUACF = [-$224,600 +$187,000(P/F,14%,40) +$56,250(P/F,14%,1)
+$37,346(P/F,14%,2) +$40,090(P/F,14%,3) +$43,510
(P/F,14%,4) +$46,687(P/F,14%,5) +$95,991(P/A,14%,5)
(P/F,14%.5) -$111,000(P/F,14%,10) +$95,991(P/A,14%,10)

20-13 Cash Flow Table

Yr.	A	B	C	D	E	F	G
0	-$490,000 investment					+$15,400)	-$224,600
0	+ 250,000 loan					)	
0	- 70,000 moving expense (applicable to yr. 1)						
1	+ 136,250	-$50,000	-$30,000	-$21250	+$15,000	0*	+$56,250
2	"	"	- 24,000	"	+ 91,000	-$24,904	+ 37,346
3	"	"	- 18,000	"	+ 97,000	- 28,160**	+ 40,090
4	"	"	- 12,000	"	+103,000	- 30,740	+ 43,510
5	"	"	- 6,000	"	+109,000	- 33,563	+ 46,687
6-10	+ 128,893	0	0	"	+107,643	- 32,902	+ 95,991
10	- 125,000					+ 14,000	-111,000
11-20	+ 128,893			- "	"	"	+ 95,991
20	- 175,000					+ 19,000	-156,000
21-30	+ 128,893			"	"	- 32,902	+ 95,991
30	- 125,000					+ 14,000	-111,000
31-40	+ 128,893			"	"	- 32,902	+ 95,991
40	+ 15,000 (equipment)						+ 15,000
40	+ 200,000 (land)					- 27,600+	+172,400

* Income tax, yr. 1 = 0.03($15,000) +0.17($14,550) - $2,924 ITC carried forward = 0

** Remaining ITC of $676 carried forward to yr. 2

\+ Capital gains tax = ($200,000 - $50,000)(0.40)(0.46) = $27,600

At i = 25%, PW = +$9,884. At i = 30%, PW = -$35,919. By interpolation, i = 26.1%

20-15 continued:

(P/F,14%,10) -$156,000(P/F,14%,20) +$95,000(P/A,14%,10)
(P/F,14%,20) -$111,000(P/F,14%,30) +$95,991(P/A,14%,10)
(P/F,14%,30)](A/P,14%,40) =+$33,855

The after-tax cash flow in the proposed plant will earn $33,855 more each year for 40 years than is necessary to earn 14% after-tax on the equity in the company.

20-16

(a) Before-Tax Cash Flows for Bonds at 10% and 6%

Year	10%	6%	Difference
0	+$500,000	+$500,000	0
1-40	- 25,000	- 15,000	+$10,000
40	- 500,000	- 500,000	0

t_e= 0.08 +(1 -0.08)(0.46) = 0.5032 or 50.32%

After-tax difference = $10,000(1 - 0.5032) per half year
= $4,968

PW of subsudy = $4,968(P/A,i%,40) = $4,968(P/A,6%,40)
= $74,750

Since the company is a regulated public utility the customers will pay less for the service by about $4,968 a year than if the company had to pay 10%, compounded semiannually for the necessary capital. The stock holders will benefit to the extent that the fair rate of return on the total investment is greater than the beforetax cost of capital included in the rate base.

20-17 Control System:

Control System:	A	B	C
Total annual disbursements	-$126,400	-$531,900	-$182,000
Depreciation	- 64,100	- 90,700	- 53,800
Total deductible expenses	-$190,500	-$622,600	-$235,800
Extra taxable income over A		- 432,100	- 45,300
Extra income tax over A		+ 233,334	+ 24,462
Cash flow after taxes	-$126,400	-$298,566	-$157,538
Capital recovery	- 130,572[a]	- 135,170[b]	- 109,591
EUAC	-$256,972	-$433,736	-$267,129

a= P(A/P,8%,20) b = P(A/P,8%,10)

(b) Only change is in capital recovery: i=14%

EUAC(A)= -$319,964; EUAC(B)= -$472,736; EUAC(C)= -$319,999

20-17 continued:

(c) If all three systems experienced the same variation in all costs (first cost as well as annual disbursements, there would be no change in the relative attractiveness of the three systems.

On the other hand, if the costs of B should be lower than the estimates by as much as 30% and the costs of A should be higher by about 20%, system B would be more attractive.

It is likely that the estimates of the first cost, installed, are likely to be more accurate than the estimates of operating disbursements. This is a good place to use the three-estimate method of examining the sensitivity of the solutions to errors in the estimates.Pessimistic, expected and optimistic estimates can be made for all the factors and the results compared.

CHAPTER 21

Capital Budgeting

General Notes

This chapter gives the teacher the opportunity to clarify some of the crude methods of analysis commonly used, and to explain in more detail how capital budgeting can make effective use of the prospective rate of return on project proposals, either mutually exclusive projects or competing projects. It is also a good place to develop the concepts of "resource allocation" in an overall development program.

21-1

(a) Crude Payout

	Proposal F	Proposal G
\$20,000/\$6,000	3.33yrs.	
\$20,000/\$4,000		5.0 yrs.

(b) Rate of return:

(A/P,i%,4)= \$6,000/\$20,000= 7.71% Proposal F

(A/P,i%,10) = \$4,000/\$20,000 = 15.10% Proposal G

(c) Difference in years 1 through 4 = -\$2,000

Difference in years 4- 10 = +\$4,000

PW = 0 = -\$2,000(P/A,i%,4) +\$4,000(P/A,i%,6)(P/F,i%,4)

by interpolation between 25 and 30%, i = 25.7%

This shows that the crude payout gives an inverse solution.

21-2

(a) Crude payout:

	Proposal L	Proposal L
\$16,000/\$5,000	3.20 yrs.	
\$16,000/\$3,500		4.57 yrs.

(b) Rate of return:

(A/P,i%,5)= \$5,000/\$16,000= 17% proposal M

i = \$3,500/\$16,000 = 21.9% proposal L

Crude payout does not take the salvage value into consideration, with the result that it indicated the less desirable proposal as the one to select.

21-3	C.F.B.T		C.F.A.T		Difference
Yr.	L	M	L	M	M-L
0	-$16,000	-$16,000	-$16,000	-$16,000	0
1	+ 5,000	+ 3,500	+ 10,500	+ 1,750	-$8,750
2	+ 5,000	+ 3,500	+ 2,500	+ 1,750	- 750
3	+ 5,000	+ 3,500	+ 2,500	+ 1,750	- 750
4	+ 5,000	+ 3,500	+ 2,500	+ 1,750	- 750
5	+ 5,000	+ 19,500	+ 2,500	+ 17,750	+15,250

Crude Payout: L = 3.2 yrs.; M = 5.0 yrs. until investment is recovered.

Rate of return: L: PW = 0 = -$16,000 +$8,000(P/F,i%,1) +$2,500(P/A,i%,5); i = 12.6%
M: i = $1,750/$16,000 = 10.9%

Incremental on M-L: PW = 0 = -$8,750 -$750(P/A,i%,3) +$15,250(P/F,i%,4); i = 9.4%

By the crude payout the proposal L is favored. While the rate of return on L is greater than on M, it is necessary to consider the rate of return on the increment before a choice between M and L can be made. If the company's i* is less than 9.4%, then proposal M should be selected. If it is between 9.4% and 12.6%, then L should be selected. If it is greater than 12.6%, neither should be selected.

21.4

yr.	C.F.B.T.	Proposal R	Proposal S
0		-$12,000	-$12,000
1		+ 5,500	+ 1,530
2		+ 4,544	+ 2.580
3		+ 3,588	+ 3,360
4		+ 2,632	+ 4,680
5		+ 1,676	+ 5,730
6		+ 720	+ 6,780
Crude Payout		2+ years	4- years

Rate of return:

R: PW= 0 = -$12,000 +$5,500(P/A,i%,6) - $956(P/G,i%,6)
i = 20%

S: PW= 0 = -$12,000 +$1,530(P/A,i%,6) -$1,050(P/G,i%,6)
i = 20%

In this case, the rates of return are identical. Since R has the shorter payout period, it is preferable.This conclusion could have been reached simply by observing that the cash flow is greater in the early years under Proposal R. Both conclusions assume that quick recovery of cash is desirable.

21-5 Proj.	Invest-ment	Annual Savings	Gross RoR	Alt.	Incremental Analysis Invest.	Savings	RoR
A1	$20,000	$4,000	15.1%				
A2	28,000	8,100	26.1%				
A3	34,000	9,300	24.2%	A3/A2	$6,000	$1,200	15.1%
B1	12,000	3,420	25.6%				
B2	16,000	4,380	24.3%	B2/B1	4,000	960	20.2%
C1	16,000	2,640	10.3%				
C2	21,000	4,620	17.7%				
D1	12,000	2,400	15.1%				
D2	16,000	4,160	22.6%				
D3	19,000	4,760	21.5%	D3/D2	3,000	600	15.1%
D4	23,000	5,980	22.6%	D4/D2	7,000	1,820	22.6%
E1	22,000	7,260	30.7%				
E2	30,000	9,300	28.5%	E2/E1	8,000	2,040	22.0%

Rankings by Rate of Return:

Alternative	Investment	RoR	Total Investment
E1	$22,000	30.7%	$22,000
A2	28,000	26.1%	50,000
B1	12,000	25.6%	62,000
D2	16,000	22.6%	78,000
D4-D2	7,000	22.6%	85,000
E2-E1	8,000	22.0%	93,000
B2-B1	4,000	20.2%	97,000
C2	21,000	17.7%	118,000
A3-A2	6,000	15.1	124,000

(a) With a MARR of 20%, the budget would contain Projects A2, B2,D4, and E2 for a total investment of $97,000.

(b) With a budget constraint of $85,000, the budget would contain Projects A2, B1, D4, and E1. The implied MARR is 22.6%. If the long range prospects are that budget constraints will effectively limit investment, the company should either consider loosening its borrowing restrictions or increasing its MARR before taxes.

21-6

In order to analyze the value of the lease, it is necessary to evaluate the potential courses of action with and without the lease. With the lease, project E2-E1 may be undertaken. The cash flow differences are as follows :

(continued on next page)

21-6 Continued:

Yr.	A A2 w/o lease	B A2 with lease	C Increment (E2-E1)	D w/ lease B+C	E Difference D-A
0	-$28,000	-$20,000	-$8,000	-$28,000	$ 0
1-4	+ 8,100	+ 5,100	+2,040	+ 7,140	-960
5-9	+ 8,100	+ 6,100	+2,040	+ 8,140	+ 40
10	+ 8,100	+ 8,100	+2,040	+10,140	+2,040
	+$53,000			+51,400	-$1,600

The lease venture will show a gross loss (PW @0%) of $1,600 even though $8,000 investment capital would be freeed to invest in a project increment (E2-E1) promising a 22.0% return. The f act that $4,000 for the first year lease would come from operating capital does not alter the fact it is cash and must be considered as a drain on the company's capital resources.

21-7

(a) Crude Payout:

	Proposal A	Proposal B	Proposal C
$20,000/$7,000=	2.86 yrs.		
$20,000/$4,000=		5.0 yrs	
$20,000/$2,600=			7.69 yrs.

(b) Rate of return:

(A/P,i%,4)=$7,000/$20,000
i = 14.96%

(A/P,i%,5)=$4,000/$20,000; i = 0%

(A/P,i%,20)= $2,600/$20,000; i = 11.54%

In this example, the crude payout method ranks the Proposal A correctly, but reverses the ranking of the other two. This and the previous problems involving crude payout prove that in spite of the fact that occasionally gives the correct answer, it cannot be relied upon to give the correct answer consistently.

21-8

Crude payout after depreciation	Proposal F	Proposal G
$20,000/($6,000-$5,000)=	20 yrs.	
$20,000/($4,000-$2,000)=		10 yrs.

Long-lived proposals show up much better than short-lived proposals because the annual depreciation is so much smaller. For F, this method gives a payout period of 20 years instead of 3.33, while G gives only 10 years compared to 5. This method really double counts for capital recovery.

21-9

The profitability index form is shown on the next page.

21-9 Continued

PROFITABILITY INDEX FORM
End-of-year Convention

Proposal ____ Description: Problem 21-9 (000 omitted)

T	0%		10%		25%		40%
-3		1.33		1.95		2.74	
-2	− 400	1.21	− 484	1.56	− 624	1.96	− 784
-1	− 1,200	1.10	− 1,320	1.25	− 1,500	1.40	− 1,680
0	− 1,350	1.0	− 1,350	1.0	− 1350	1.0	− 1,350
+1		.909		.800		.714	
Total	− 2,950		− 3,154		− 3,474		− 3,814
		Receipts or Cost Reductions					
1	+ 150	.909	+ 136	.800	+ 120	.714	+ 107
2	+ 225	.826	+ 186	.640	+ 144	.510	+ 115
3	+ 375	.751	+ 282	.512	+ 192	.364	+ 137
4	"	.683	+ 256	.410	+ 154	.260	+ 98
5	"	.621	+ 233	.328	+ 123	.186	+ 70
6	"	.565	+ 212	.262	+ 98	.133	+ 50
7	"	.513	+ 192	.210	+ 79	.095	+ 37
8	"	.467	+ 175	.168	+ 63	.068	+ 25
9	"	.424	+ 159	.134	+ 50	.048	+ 18
10	"	.386	+ 145	.107	+ 40	.035	+ 13
11	"	.351	+ 132	.086	+ 32	.025	+ 9
12	"	.319	+ 120	.069	+ 26	.018	+ 7
13	"	.290	+ 109	.055	+ 21	.013	5
14	+ 275	.263	+ 72	.044	+ 12	.009	2
15	+ 875	.239	+ 209	.035	+ 31	.006	5
Total	+ 5,650		+ 2,618		+ 1,185		+ 698
Ratio	0.522		1.204		2.931		5.464

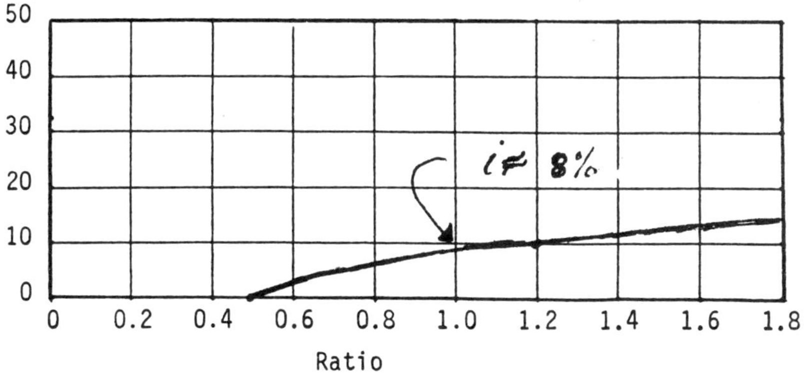

21-9 Continued

Find i so that PW= 0 = -$400(F/P,i%,2) -$1,200(F/P,i%,1) -$1,350 + $150(P/F,i%,1) + $225(P/F,i%,2) +$375(P/A,i%,11) (P/F,i%,2) + $275(P/F,i%,14) + $875(P/F,i%,15)

Interpolation between 7% and 8%, indicates i = 7.3%

21-10

Change the affected lines on the Profitability Index Form for Problem 21-9 as follows:

T	0%	10%	25%	40%
-2	-$ 200	-$ 242	-$312	-$ 392
Total	-$2,750	-$2,912	-$3,162	-$3,422
15	+ $819	+$ 196	+$ 29	+$ 5
Total	+$5,594	+$2,605	+$1,813	+$ 698
Ratio	0.491	1.117	1.74	4.9

Unless plotted on a large scale chart, this method will not show any real difference in the estimate of the i for the project.

Find i so that PW= 0 = -$200(F/P,i%,2) -$1,200(F/P,i%,1) -$1,350 +$150(P/F,i%,1) +$225(P/F,i%,2) +$375(P/A,i%,11)(P/F,i%,2) +$275(P/F,i%,14) +$819(P/F,i%,15)
i is approximately 8%.

21-11 Capital available = $250,000; i* = 18% before income tax

Project	Investment	Annual Savings	RoR	Incremental Ror	Accept or Reject
A1	$50,000	$12,00	20.18%		Acceptable
A2	100,000	25,000	21.48		"
A3	125,000	27,000	17.17		Reject
B1	60,000	14,800	21.0		Accept
B2	80,000	16,500	15.9		Reject
C1	40,000	8,000	15.1		"
C2	50,000	12,000	20.2		Accept
C3	75,000	16,000	16.8	9.6%on C3/C2	Reject
C4	85,000	20,000	19.6	18.76% on C4/C2	Accept
D1	30,000	7,000	19.4		"
D2	38,000	8,000	16.5		Reject
D3	47,000	11,000	19.4	19.4 on D3/D1	Accept

21-11 Continued

E1	$43,000	$10,000	19.3%		Accept
E2	60,000	14,500	20.4		"
E3	70,000	15,700	18.2	3.5% E3/E2	Reject

(a)

Proj.	Investment	RoR	Cummulative Investment
A2	$100,000	21.4$	$100,000
B1	60,000	21.0%	160,000
E2	60,000	20.4%	220,000
C2	50,000	20.2%	270,000
C4-C2	35,000	18.7%	305,000
D1	30,000	19.4%	335,000
D3-D1	17,000	19.4%	352,000

(b) Without borrowing, the projects selected would be A2, B1, E2, and D1, for a total of $250,000

(c) With borrowing, the projects selected would be A2, B1, E2, D3, and C4 for a total investment of $352,000, of which $102,000 would be borrowed at 14%.

21-12

With continuous compounding the annual receipts are assumed to be spread uniformily over the entire year. Therefore, the company gets the money earlier and the present worth is greater than with the end-of-year convention. It is to be expected that the prospective rate of return shown by the profitability index chart will be great than by the end-of-year convention.

Profitability index form is on next page.

21-13 Proposal A: n = 6; P = -$18,000; A = $5,000; G= -$300; = 0.

Payout by successive years: $5,000+ $4,700 +$4,400 +$4,100 = $18,200. Thus, payout is a little less than 4 years.

Find i so that PW = 0 = -$18,000 +$5,000(P/A,i,6) -$300(P/G,i%,6) ; By interpolation, i = 11.75%

Proposal B: n = 8; P = -$18,000; A = $3,500; L = +$4,000. Payout = $18,000/$3,500 = 5.14 years

Find i so that PW = 0 = -$18,000 +$3,500(P/A,i%,8) +$4,000(P/F,i%,8); i = 13.45%

Neither proposal meets both criteria. Clearly Proposal B is more attractive than A, but it may be rejected if the payout criterion is deemed to be more significant than the rate of return. Any significant salvage value is ignored in the payout method but may be very important in the rate of return method of analysis.

21-12 continued

PROFITABILITY INDEX FORM
Continuous Compounding

Proposal K2 Description: Problem 21-12

T	0%		10%		25%		40%
-3		1.29		1.74		2.30	
-2		1.53		1.39		1.65	
-1		1.05		1.12		1.18	
0	- 80,000	1.0	- 80,000	1.0	- 80,000	1.0	- 80,000
+1		.954		.896		.849	
Total	- 80,000		- 80,000		- 80,00		- 80,000
		Receipts or Cost Reductions					
1	+ 22,000	.954	+ 20,988	.896	+ 19,712	.849	+ 18,678
2	+ 23,800	.867	+ 20,634	.717	+ 17065	.607	+ 14447
3	+ 25,600	.788	+ 20,173	.574	+ 14,694	.433	+ 11,085
4	+ 25,800	.717	+ 18,498	.459	+ 11,842	.310	+ 7,998
5	+ 26,000	.652	+ 16,952	.367	+ 9,542	.221	+ 5,746
6	+ 25,000	.592	+ 14800	.294	+ 7,350	.158	+ 3,950
7	+ 20,000	.538	+ 10,760	.235	+ 4,700	.113	+ 2260
8	+ 36,000	.490	+ 17,640	.188	+ 6,768	.081	+ 2916
9		.445		.150		.058	
10		.405		.120		.041	
11		.368		.077		.029	
12		.334		.062		.021	
13		.304		.049		.015	
14		.276		.039		.011	
15		.251		.032		.008	
Total	+ 204,200		140,445		+ 91,673		+ 67,080
Ratio	0.39		0.569		0.873		1.193

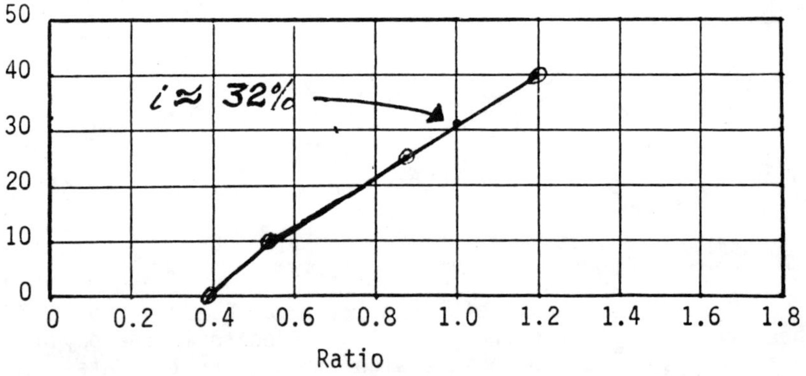

21-14

PROFITABILITY INDEX FORM
Continuous Compounding

Proposal ____ Description: *Problem 21-14*

T	0%		10%		25%		40%
-3		1.29		1.74		2.30	
-2	− 400	1.15	− 460	1.39	− 556	1.65	− 660
-1	− 1,200	1.05	− 1260	1.12	− 1344	1.18	− 1416
0	− 1,350	1.0	− 1350	1.0	− 1350	1.0	− 1350
+1		.954		.896		.849	
Total	− 2950		− 3070		− 3250		− 3426
		Receipts or Cost Reductions					
1	+ 150	.954	+ 143	.896	+ 134	.849	+ 127
2	+ 225	.867	195	.717	161	.607	137
3	+ 375	.788	296	.574	215	.433	162
4	"	.717	269	.459	172	.310	116
5	"	.652	245	.367	138	.221	83
6	"	.592	222	.294	110	.158	59
7	"	.538	202	.235	88	.113	42
8	"	.490	184	.188	71	.081	30
9	"	.445	167	.150	56	.058	22
10	"	.405	152	.120	45	.041	15
11	"	.368	138	.077	29	.029	11
12	"	.334	125	.062	23	.021	8
13	"	.304	114	.049	18	.015	6
14	+ 275	.276	76	.039	11	.011	3
15	+ 875	.251	220	.032	28	.008	7
Total	+ 5,650		2748		1299		828
Ratio	0,52		1.117		2.50		4.14

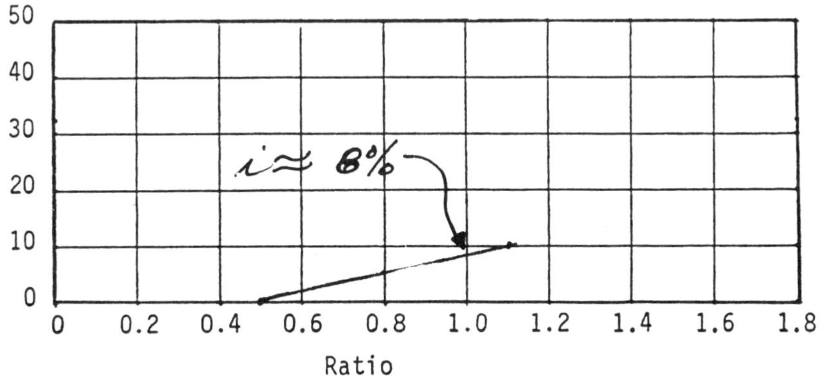

21-14 Continued

Continuous compounding increases the prospective rate of return slightly, but with the profitability index chart, it is difficult to detect how much. The problem can be expanded by asking the students to compute the rate of return with continuous compounding.

21-15

No solution can be given for this suggested term project report.

APPENDIX A

Continuous Compounding of Interest and the Uniform-Flow Convention

General Notes

The wide spread use of computers in economic analysis has led to much wider use of the uniform-flow convention (or continuous compounding) because it is much easier to program the computer to perform the required computation than is the end-of-year convention, using $(1 + i)^n$. This is also true of certain programmable hand-held calculators. The students should be aware of this advantage and understand its usefulness as well as the limitations that should be applied. It is not appropriate to assume uniform-flow for all cash flows. See Problem A-3.

Most students are familiar with open end charge accounts, borrowing from banks, financing cars, and savings accounts, yet few have ever considered the true cost of borrowed money, or the true rate of return on their savings accounts. The personal approach will help get the students interested in the concepts of continuous compounding and the uniform-flow concept.

Many business firms tend to pay the actual expenses more or less uniformly throughout the year and to collect their revenues fairly uniformly. In that case, the uniform-flow convention is a better representation of the real world situation than the assumption of year-end payments and receipts. Unfortunately, in engineering economy studies, most of the data employed are estimates of future events and the estimates probably will be in error by a much larger degree than the difference between the year-end convention and the uniform-flow convention. One should guard against assuming greater confidence in the resulting analysis because the uniform-flow has been used.

A-1 (a) let $r = 0.0575$; $e^r - 1 = 1.059185 - 1$

$i_{eff.} = 5.9185\%$

(b) $i = (1 + r/365)^{365} - 1 = 0.059180\%$

$i_{eff.} = 5.9180\%$;

The difference between continuous and daily compounding only occurs in the 6th decimal place when the interest is expressed as a decimal fraction.

A-2 (a) $i = (1 + 0.015)^{12} - 1 = 19.562\%$

(b) $i = e^{r} - 1 = e^{0.18} - 1 = 19.722\%$

(a) or 19.562%, is the effective annual rate of 1.5%/month, compounded monthly.

A-3 Annual receipts and disbursements:

Receipts:	52($185)=	$9,620
Disbursements:	-12($520)=	- 6,240
Annual taxes:		- 240
Net cash flow =		+$3,140

(a) End-of-year convention:
PW= -$6,000 +$3,140(P/A,25%,6) = +$3,266.14

(b) Uniform-flow convention:
$PW = -\$6,000 + \$3,380(P/\bar{A},25\%,6) - \$240(P/A,25\%,6)$
$= -\$6,000 + \$3,380(3.307) - \$240(2.951) = +\$4,469.42$

(c) Percentage difference : ($4,469.42-$3,266.14)/$3,266.14 = 0.3683 or 36.83%. A very substantial difference, due, primarily to the high interest rate required before income tax.

A-4

		Difference in Interest
(a)	$\$1,000(1.12) = \$1,120$	
(b)	$\$1,000(1 + 0.06)^{2} = \$1,123.60$	3.60%
(c)	$\$1,000(1 + 0.03)^{4} = \$1,125.50$	1.91%
(d)	$\$1,000(1 + 0.01)^{12} = \$1,126.83$	1.32%
(e)	$\$1,000(1 + 0.12/52)^{52} = \$1,127.34$	0.51%
(f)	$\$1,000(1 + 0.12/365)^{365} = \$1,127.47$	0.13%
(g)	$\$1,000(e^{0.12}) = \$1,127.50$	0.03%

As the frequency of compounding increases the rate of increase in the effective annual interest decreases. Going from annual compounding to monthly compounding increases the effective rate by 6.83%, but from monthly to continuous compounding only increases the rate by 0.67%.

A-5 (a) Uniform-flow convention:

Proj. G: $PW = 0 = -\$20,000 + \$9,200(P/\bar{A},i\%,3)$

$(P/\bar{A},i\%,3) = \$20,000/\$9,200 = 2.174$

by interpolation, i = 25.59%

A-5 Continued

Proj.H: $PW = 0 = -\$20,000 + \$3,700(P/\bar{A},i\%,25)$

$(P/\bar{A},i\%,25) = 5.4054$; $i = 20.11\%$

(b) End-of-year convention:

Proj. G: $(P/A,i\%,3) = \$20,000/\$9,200 = 2.1739$

$i = 18.01\%$

Proj. H: $(P/A,i\%,25) = \$20,000/\$3,700 = 5.4054$

$i = 18.22\%$

A-6 (a) Uniform-flow convention:

Proposal M: $PW=0=-\$75,000+\$17,200(P/\bar{A},i\%,6)$

$(P/\bar{A},i\%,6) = 4.3605$; $i = 11.93\%$

Proposal N: $PW=0=-\$75,000 + \$11,700(P/\bar{A},i\%,10) + \$15,000(P/F,i\%,10)$

By interpolation between 10% and 12%, $i = 11.95\%$

(b) End-of-year convention:

Proposal M: $PW=0= -\$75,000 + \$17,200(P/A,i\%,6)$

$(P/A,i\%,6) = 6.4054$; $i = 9.96\%$

Proposal N: $PW=0=-\$75,000 + \$11,700(P/A,i\%,10) + \$15,000(P/F,i\%,10)$; $i = 10.76\%$

A-7 Proposal M : There will be no change in the prospective rate of return by eaither method.

Proposal N : $PW = 0 = -\$75,000/(P/\bar{A},i\%,1) + \$11,700 (P/F,i\%,10) + \$15,000(P/F,i\%,10)$

@ $i = 10\%$, $PW = +\$1,255.04$

@ $i = 12\%$, $PW = -\$4,500.40$; $i = 10.44\%$

No changes will be made in the rates of return by the end-of-year method.

The fact that the machine involved in Proposal N will be built over the year preceeding operations increases the first cost using the uniform-flow convention results in an i= 10.44% instead of 11.95% obtained by the end-of-year convention.

A-8 (a) $i = (1 + 0.0525/365)^{365} - 1 = 5.3899\%$

(b) $i = e^{0.0525} - 1 = 5.3903\%$

(c) Deposits can be made or funds can be withdrawn at any time during the working day. Daily compounding eliminates arguments about interest due and continuous compounding is not well understood by the general public.

A-9 End-of-year convention : (000 omitted)

$PW = 0 = -\$20,000(F/P,i\%,1) - \$10,000 + \$1,700(P/A,i\%,40$
$+\$100(P/G,i\%,40)$

@ $i = 8\%$, $PW = +\$933$

@ $i = 10\%$, $PW = -\$6,480$; $i = 8.13\%$

Uniform-flow convention: (000 omitted)

No tables are given for converting \$100,000 gradient, flowing uniformly throughout the year, to a PW. Thus, the \$100,000 can be converted to a end-of-year gradient by $\$100,000/(P/\overline{F},i\%, 0\text{-}1)$. Then the end-of-year factors can be used.

$$PW = 0 = -\$20,000/(P/\overline{F},i\%,1\text{-}2) - \$10,000/(P/\overline{F},i\%,0\text{-}1)$$
$$+\$1,700(P/\overline{A},i\%,40) + \$100(P/G,i\%,40)/(P/\overline{F},i\%,0\text{-}1)$$

@ $i = 8\%$, $PW = +\$1,336$

@ $i = 10\%$, $PW = -\$6,782$; $i = 8.16\%$

With very large amount of money involved, the use of the uniform-flow convention can make some real differences. In this case, the uniform-flow assumption is more realistic than the end-of-year assumption, but it turns out that it makes very little difference in the results obtained. If a higher interest rate had been used (required) the difference would have been greater.

A-10 Analysis of anticipated changes if the uniform-flow convention is used :

First cost will be increased; capital recovery increased.

Present worth of all annual disbursements except property taxes and insurance will be increased.

Net result will be to increase the EUAC of each of the three alternatives.

Assume the savings in income tax for B and C will be spread uniformly throughout the year.

$$EUAC(A) = -\$1,282,000(A/P,8\%,20)/(P/\overline{F},8\%,0\text{-}1)$$
$$-\$75,200(P/\overline{A},8\%,20)(A/P,8\%,20) - \$51,200 = \underline{-\$265,032}$$

$$EUAC(B) = -\$907,000(A/P,8\%,10)/(P/\overline{F},8\%,0\text{-}1)$$
$$-\$262,226(P/\overline{A},8\%,10)(A/P,8\%,10) - \$36,300 = \underline{-\$449,315}$$

$$EUAC(C) = -\$1,076,000(A/P,8\%,20)/(P/\overline{F},8\%,0\text{-}1)$$
$$-\$114,338(P/\overline{A},8\%,20)(A/P,8\%,20) - \$43,200 = \underline{-\$275,917}$$

Each of these EUA costs is higher than those in Problem 20-17.

A-11 By end-of-year convention, revenue requirement

$= \$1,000,000/\$5,000,000 = .20$ or 20% of Rate Base

By uniform-flow method, rate base becomes

$= \$2,500,000/(P/\overline{F},10\%,0\text{-}1) + \$2,500,000/(P/\overline{F},10\%,1\text{-}2)$

$= \$2,671,511 + \$2,883,173 = \$5,554,684$

A-11 Continued

If revenue requirement is based on uniform-flow:

with no change in rate base: Revenue Requirement
$= \$1,000,000(P/\bar{F},10\%,0\text{-}1) = \$1,000,000(0.9538) = \$953,800$

with rate base computed by uniform-flow method:
Revenue Requirement $= (0.20)(\$5,554,684)(P/\bar{F},10\%,0\text{-}1)$
$= \$1,059,612$

APPENDIX B

The Analysis of Prospective Cash Flow Series That Have Two or More Reversals of Sign

B-1 An analyst should first consider the purchase of the lease assuming that water flooding will not be undertaken. The cash flow series to be evaluated is :

Yr.	Cash Flow
0	-$180
1	+120
2	+ 90
3	+ 60
4	+ 30

It is evident from inspection that the prospective rate of return is relatively high. Trial and error calculations indicate that the rate is approximately 31.5%. Thus, by normal standards, the investment should be made. (Unless i* is greater than 31.5%.) Now the analyst can evaluate the attractiveness of water flooding.

Water Flooding

Yr.	Cash Flow
0	-$1,810
1	+600
2	+500
3	+400
4	+300
5	+200
6	+100

Since water flooding is not required, that possible investment can be analyzed separately. Trial and error calculations of the cash flow resulting from flooding indicate that the investment will yield only 6% rate of return. The investment in water flooding is not nearly as attractive as the original investment in the lease.

B-2 This problem changes the cash flow at time zero from 0 to -$180,000, and results in there now being three reversals of sign. It is still evident that most of the cash flow takes place i years 5 through 11. One approach would be to assume an auxiliary rate and apply it to all of the cash flows from year zero through year 4, and compute the prospective rate of the return on the investment portion of the -$1,810,000 cash flow at year 5.

If the auxiliary rate is assumed to be 0% (an extreme value), X will be computed to be +$120, and X-$1,810 = -$1,690. On that basis, the prospective rate of return is about 8.8%. If a high rate, say 15%, is chosen for the auxiliary rate, the new X will be +$98.6, and X-$1,810 = -$1,711.4. On that basis the prospective rate of return is about8.3%.

B-2 Continued

This problem has three reversals of sign and could be viewed as first an investing period, followed by a borrowing period and finally an investing period. Since the major cash flows occur in the last period, year 5 through 11, two auxiliary rates could be used. The first could be the i* on the original investment at time zero. If that is assumed to be 20%, the Value of X at year 5 based on a borrowing rate of 10%, would be +$118,000. Then the prospective rate of return on the X-$1810,000 =-$1,692,000 would be about 7.5%.

B-3 As $x = \frac{1}{1 + i}$; $i = \frac{1 - x}{x}$

It is evident that if x is negative, the numerator of the fraction (1-x)/x must be positive and the denominator must be negative; therefore i must be negative whever x is negative. For positive values of x, the denominator will be positive and the numerator will be positive if x is less than 1 and positive if x is greater than 1. Therefore, a positive value of x may give either a positive or negative value of i.

B-4 (000 omitted in all the following calculations)

Auxiliary rate = 0%: X = \$120 +\$90 +\$60 +\$ 0 = \$300
-\$1,810 + X = -\$1,510
PW = 0 = -\$1,510 +\$600(P/A,i%,6) -\$100(P/G,i%,6)
@ i = 12%, PW = +\$64; @ i = 15%, PW = -\$34; i = 14.0%

Auxiliary rate = 5%: X= \$120 -\$30(A/G,5%,5) (F/A,5%,5)=\$348
-\$1,810 + X = -\$1,462
PW = 0 = -\$1,462 +\$600(P/A,i%,6) +\$100(P/G,i%,6)
@ i = 15%, PW = +\$14; @ i = 20%, PW = -\$124; i = 15.5%

Auxiliary rate = 10%: X = \$120 -\$30(A/G,10%,5) (F/A,10%,5)
= \$401
-\$1,810 + X = -\$1,409
PW = 0 = -\$1,409 +600(P/A,i%,6) -\$100(P/G,i%,6)
@ i = 15%, PW = +\$67; @ i = 20%, PW = -\$71; i = 17.4%

B-5 (Omit final 000 in all the following calculations)

Auxiliary rate = 0%: Let year 4 be new 0 date:

Yr.	Cash Flow
0	+\$100
1 to 6	+200/yr.
7 " 16	+100/yr
17	-\$3,000

PW = 0 = \$100 +\$100(P/A,i%,16) +\$100(P/A,i%,6)
-\$3,000(P/F,i%,17)

@ i = 2.5%, PW = -\$15.3
@ i = 3%, PW = +\$82.8; i = 2.6%

Auxiliary rate = 10%:
Available positive cash flow for year 5 = -\$700(F/P,10%,5)
+\$200(F/A,10%,5) = +\$93.7

B-5 Continued

Yr.	Cash Flow
0	+$93.7
1 to 5	+200 /yr
6 " 15	+100/yr.
16	$3,000

PW = 0 = +$93.7 +$100(P/A,i%,15)
+$100(P/A,i%,5) -$3,000(P/F,i%,16)
@ i = 4%, PW = +$49.0
@ i = 3.5%, PW = -$33.2; i = 3.7%

Auxiliary rate = 20%: Let year 7 be 0 date:

Yr.	Cash Flow
0	+ $75
1 to 3	+ 200/yr.
4 " 13	+ 100/yr.
14	-3,000

Available positive cash flow for year 7 =
-$700(F/P,20%,7) +$200(F/A,20%,7) = +$75

PW = +$75 +$100(P/A,i%,13) +$100(P/A,i%,3)
-$3,000(P/F,i%,14)
@ i = 6%, PW = -$99.3
@ i = 7%, PW = -$9.8; i = 6.9%

B-6

Years	i=0%	i=2%	i=6%	i=10%	i=15%	i=20%	i=30%
0	-$18.00	-$18.00	-$18.00	-$18.00	-$18.00	-$18.00	-$18.00
1 - 5	+50.00	+ 47.13	+ 42.12	+ 37.91	+ 33.52	+ 29.91	+ 24.36
6 -10	-50.00	- 42.70	- 31.48	- 23.53	- 16.67	- 12.01	- 6.56
20	+20.00	+ 23.46	+ 6.24	+ 2.97	+ 1.22	+ 0.52	+ 0.11
Total	+ 2.00	- 0.11	- 1.12	- 0.65	+ 0.07	+ 0.42	- 0.09

B-7

Years	i=0%	i=5%	i=25%	i=50%	i=75%
0	-$ 500.00	-$ 500.00	-$ 500.00	-$ 500.00	-$ 500.00
1 - 4	+ 500.000	+1,772.98	+1,180.80	+ 802.47	+ 595.59
5	-1,600.00	-1,253.64	- 524.29	- 210.70	- 97.48
Total	-$ 100.00	+ 19.34	+ 256.51	+ 91.77	- 1.89

Two solving rates = 4.007% and 74.464%

B-8 Auxiliary rate (investing) = 0%: year 1 is new 0 date
PW = 0 = -$1,600(P/F,i%,4) +$500(P/A,i%,3); i = 3.263%

Auxiliary rate = 5%: Year 2 is new zero date:
Cash available year 2 = -$500(F/P,5%,2) +$500(F/A,5%,2)=+$
=+$473.75
PW = 0 = +$473.75 +$500(P/A,i%,2) -$1,600(P/F,i%,3); i= 4.208%

Auxiliary rate = 10%; Year 2 is new zero date:
Cash available = -$500(F/P,10%,2) +$500(F/A,10%,2)= +$445.00
PW = 0 = +$445.00 +$500(P/A,i%,2) -$1,600(P/F,i%,3); i= 5.289%
Auxiliary rate = 15%; Year 2 is new zero date:
Cash available = -$500(F/P,15%,2) +$500(F/A,15%,2) = +413.75
PW = 0 = +$413.74 +$500(P/A,i%,2) -$1,600(P/F,i%,3); i= 6.526%

The cost of borrowing is relatively insensitive to the i* for investing.

B-9 Plot the values of the PW for different interest rates found by : PW = -\$16,000 +\$50,000(P/A,i%,2) -\$100,000(P/F,i%,3)

Values are:

i%	PW	i%	PW
0%	-\$16,000	200%	+\$2,518
10	- 4,355	250	+ 35
15	- 466	275	- 1,007
50	+ 9,926	300	- 1,936
100	+ 9,000		

B-10 Year 1 is the new zero date for all auxiliary rates.

Auxiliary rate (investing) = 0%; Cash available = \$34,000

PW = 0 = +\$34,000 +\$50,000(P/F,i%,1) -\$100,000(P/F,i%,2)

i (borrowing cost) = 13.06%

Auxiliary rate = 10%; Cash available = \$50,000 -\$16,000(F/P,10%,1) = \$32,400

PW = 0 = +\$32,400 +\$50,000(P/F,i%,1) -\$100,000(P/F,i%,2)

i = 14.72%

Auxiliary rate = 15%; Cash available = \$50,000 -\$16,000(F/P,15%,1) = \$31,600

PW = 0 = +\$31,600 +\$50,000(P/F,i%,1) -\$100,000(P/F,i%,2)

i = 15.58%

B-11 Examine the Cash Flow Table for G-F in Table 7-1. The general equation for the PW can be written and then solved for different interest rates.

PW = 0 = -\$30,000 -\$2,000(P/A,i%,5) +\$50,000(P/F,i%,5)
+\$1,500(P/A,i%,3)(P/F,i%,5) -\$30,000(P/F,i%,8)
-\$1,000(P/A,i%,2)(P/F,i%,8) +\$40,000(P/F,i%,10)
+\$2,000(P/A,i%,5)(P/F,i%,10) -\$10,000(P/F,i%,15)

@ i = 7%, PW = +\$2,610; @ i = 8%, PW = +\$590
@ i = 10%, PW = -\$3,074

Since positive cash flows dominate the future cash flows, the PW of each will decrease as the i is increased. Thus, the PW for the entire series will continually decrease and no other positive value of i will give a PW = 0.

Interpolating between 8% and 10%, gives an estimated i = 8.32%

B-12 (a) PW = 0 = +\$400 -\$1,000(P/F,i%,1) +\$300(P/A,i%,3)(P/F,i%,2)
@ i = 20%, PW = +5.517; @ i = 21%, PW = - 1.490
i = 20.75%

(b) Investment at year 1 is -\$1,000 +\$400(1.14) = -\$544
PW = -\$544 + \$300(P/A,i%,3)(P/F,i%,1)
i = 18.66%

B-12 Continued

(c) Auxiliary borrowing rate = 0, Investment at time 1 = $600

PW = 0 = -600 +$300(P/A,i5,3)(P/F,i5,1)

i = 14.72%

Auxiliary rate = 20%; Investment at time 1 =
-$1,000 +$400(1.20) = -$1,000 + $480 = -$520

PW = 0 = -$520 +$300(P/A,i%,3)(P/F,i%,1)

i = 20.53%

A spread of 20% in the borrowing rate only results in a spread of about 6% in the return on the investment phase. Relatively insensitive.

B-13 (a)

i%	PW	
10%	-$1,119.73	Computer solution: Two solving
15%	+ 2,700.42	rates: i = 11.46% and i = 41.75%
20%	+ 4,380.58	
40%	+ 938.56	Solutions by interpolation between
45%	- 995.51	10% and 15% and between 40% and
50%	- 3,059.01	50% will be substantially different.

(b)This is an investing problem, because the major flows occur immediately after the primary investment. The borrowing period occurs at the end of the series, so one must work backwards from the end to find out what positive cash flows are the returns on the initial investment.

PW = -$60,000(P/F10%,1) - $50,000(P/F,10%,2) -$40,000 (P/F,10%,3) = -$125,920 at end of year 3.

$125,920 - $90,000 = $35,920 to be applied to year earlier receipt.

$50,000-$35,920(P/F,10%,1) = $17,345

PW = 0 = -$70,000 +$60,000(P/Fi%,1) +$17345(P/F,i%,2)

i = 8.54%

With this low rate of return, it is easy to see why the company resisted the government's pressure to withdraw the gas earlier. The company's i* for investments in proved fields was about 20%, after-income-tax.

B-14 (a) Taking the difference in the two cash flows, we have:

PW = 0 = -$30,000 +$70,000(P/F,i%,20) -$500(P/A,i%,20) (P/F,i%,20); i = 4.8%

This indicates that the investment in the longer-lived structure is not very attractive.

B-14 Continued

(b) The PW of this series of cash flows will become more negative as the i is increased above 4.8% because of the large +$70,000 cash flow at year 20, which will diminish rapidly as the interest rate is increased. The only other solving rate for the series is negative and meaningless.

(c) PW of the negative cash flows of $500 a year can be considered to be repayment of a sum borrowed at year 20. If a reasonable rate for borrowed money of 12 percent is used, the positive cash flow at year 20 resulting from the investment of $30,000 at year 0 is:

$70,000 -$500(P/A,12%,20) = +$66,265

PW = 0 = -$30,000 +$66,265(P/F,i%,20); i = 4.0%

Clearly this is not an attractive investment.

B-15 (a) The second solving rate of:

PW = 0 = +$100 -$200(P/A,i%,5) +$1,000(P/F,i%,5)

i = 6.4%

(b) Repayment of the first receipt of $100 at year 1 will require $100(1.12) = $112.-$200+$112 =-$88

PW = 0 = -$88 -$200(P/A,i%,3) +$800(P/F,i%,4)

i = 7.0% which is less than the i* of 15%

Proposal should be rejected.

APPENDIX C

The Reinvestment Fallacy in Project Evaluation

C-1

(a) ROR after income taxes:
PW = 0 = -\$75,700 + \$20,000(P/A,i%,20)
-\$1,000(P/G,i%,20)
by interpolation, i = 20.0%

(b) Compound amount of the positive cash flow in 20 years at 3%:
[\$20,000(P/A,3%,20)-\$1,000(P/G,3%,20)](F/P,3%,20)= \$308,375
At what interest rate must \$75,700 now be compounded to amount to \$308,375 in 20 years?
(F/P,i%,20) = \$308,375/\$75,700 = 4.074; i = 7.3%

(c) The i found in (a) is the interest rate at which the original investment will be returned to the investor, but in (b), the i of 7.3% is a combination of the 20.0% earned on the project and 3% earned elsewhere on the cash throw-off from the initial investment.

C-2

(a) ROR after income taxes:
PW = 0 = -\$170,750+\$20,000(P/A,i%,20)-\$1,000(P/G,i%,20)
i = 3.0%

(b) Compound amount of positive cash flow when invested at 12%:
=[\$20,000(P/A,12%,20)-\$1,000(P/G,12%,20)](F/P,12%,20)
= \$1,007,189
(F/P,i%,20) = \$1,007,189/\$170,750 = 5.899
i = 9.2%

(c) See part (c) of Problem C-1. 9.2% is a combination of 3% and 12% separate investments.

C-3

(a) As (A/P,20%,2) is 0.655, Project A clearly has a 20% rate of return. PW of B series is $106,450 at 30% and $98,712 at 35%; interpolation gives 34%.

(b) For Project A, F on date 2 with compounding at 4% = $133,620; interpolation for F/P of 1.3362 for 2 years gives a rate of return of 15.6%. For Project B, F on date 10 is $275,012; interpolation for an F/P of 2.750 for 10 years gives a rate of return of 10.6%.

(c) The use of the estimated terminal dates of the respective series of cash flows for different projects results in different percentages of dilution of project rate of return by the reinvestment rate of return. Thus the 10 year Project B that yields 34% appears to yield only 10.6% because of the dominance of the 4% reinvestment return in the analysis method used by the WXY Company.

C-4

(a) PW of positive cash flows is $100,000 at 3-1/2%.

(b) At 15%, the compound amount at date 20 of the positive cash flows from years 1 to 20 is $20,000(6.742)(8.137) +$1,000(47.58) = $1,097,200+$47,600 = $1,144,800. Thus F/P = 11.448. Interpolation between F/P factors for 20 years and 12% and 15% indicates a rate of return of 12.8%.

C-5

(a) (F/P,i%,5) = 1.70. Interpolation indicates that i is 11.2%.

(b) Because all positive cash flow is at terminal date (date 5), no reinvestment at 15% enters into the calculation and i still appears to be 11.2%.

(c) No. The reason why Project E appears so good with this analysis method is that the 15% assumed reinvestment rate carries so much weight in the conclusion as to the project's merit. Even though the project itself will only yield 3.5%, the concentration of most of the positive cash flow in the early years in combination with the distant terminal date causes the computed rate of return to be much closer to the 15% than to the 3.5%.

C-6

True rate of return on Q obviously is 50%.
For R, (F/P,i%,20) = 3.50 and rate of return is about 6.4%.

But if the XYW Company's system is used with a reinvestment at 4%, for Q, F_{20} = \$150,000(2.107) = \$316,000; (F/P,i%,20) = 3.16 and rate of return appears to be about 5.9%, less than that of R.

The reinvestment assumption will have no influence on R because there is nothing to reinvest. However, the reinvestment assumption reduces the return from Q from 50% to 5.9% because it creates a mixture of 50% investment for one year with a 4% investment for the subsequent 19 years.

Project S obviously has a negative rate of return; half the invested funds are lost in one year, Project T has a value of (F/P,i%,20) of 6.80 for which i is a little over 10%. But if the XYW Company's system is used with a reinvestment at 15% for S, F_{20} = \$50,000(14.232) = \$711,600; (F/P,i%,20) = 7.116 and the rate of return appears to be 10.3%, more than that of T.

The reinvestment assumption will have no influence on T because there is nothing to reinvest. However, the assumption of a reinvestment at 15% for 19 years of the proceeds from the unsuccessful investment in S actually makes S better than T.

The foregoing example brings out the point that the difficulties in making sound comparisons using a reinvestment assumption are not corrected by the choice of a common terminal date for all proposals.

C-7

(a) The rate of return is found from:
NPW = 0=-\$10,000+\$2,800(P/A,i%,5)+\$2,000(P/F,i%,5)
i = 16.5%
(b) Net annual return = \$5,000-\$2,200-(\$10,000-\$2,000) (A/F,10%,5) = \$1,489.60
Rate of Return = \$1,489.60/\$10,000 = 14.9%
(c) Net annual return = \$5,000 - \$2,200 -\$8,000(A/F,6%,5) = \$1,380.80
Rate of Return = \$1,380.80/\$10,000 = 13.8%
(d) Net annual return = \$5,000 - \$2,200 -\$8,000(A/F,16.5%,5) = \$1,647.48
Rate of Return = \$1,647.48/\$10,000 = 16.5%

C-7 continued

While an interesting mathematical exercise, the argument is irrelevant. Depreciation, by any method, is not a cash flow. If invested capital is recovered, it is simply because the excess of receipts over disbursements, after consideration of income taxes, meets or exceeds the amount of capital invested. If this proposal were to be compared with another on a mutually exclusive basis, the cash flow differences between them would determine the choice on the same basis.

C-8

(a) The rate of return is found from:
NPW = 0 = -$10,000+$1,600(P/A,i%,5)+$2,000(P/F,i%,5)
i = 0% [5($1,600)+$2,000 = $10,000]

(b) Net annual return = $3,800 - $2,200
-$8,000(A/F,10%,5) = $289.60
Rate of Return = $289.60/$10,000 = 2.9%

(c) The cash throw-offs are compound forward to time 5 as follows:
$2,000+$1,600(F/A,10%,5) = $2,000+$1,600(6.105) = $11,768
Rate of Return = (P/F,i%,5) = $10,000/$11,768 = 0.8498
i = 3.3%

(d) Since the project only returns the original investment, the return is explainable only as the result of the reinvestment of funds elsewhere at 10%. In (b), only the recovery of the original capital investment is assumed to be reinvested; in (c), all cash throw-offs are assumed to be reinvested. Neither method measures the worth of the original investment proposal.

C-9

(a) Annual net return = $4,200-$2,200-$10,000(A/F,6%,10)
= $1,241.30
Rate of Return = $1,241.30/$10,000 = 12.4%
This would not be an attractive investment based on a company standard of 15%.

(b) The rate of return on this proposal is found from:
NPW = 0 = -$10,000+$2,000(P/A,i%,10)
i = 15.1%, therefore the proposal does meet the company's standard of attractiveness.

C-9 continued

(c) Annual net return = \$4,200-\$2,200-\$10,000(A/F,15%,10) = \$1,507.5
Rate of Return = \$1,507.5/\$10,000 = 15.1%
This result is virtually the same as that in (b) because i* is so close to the true rate of return.

(d) NPW = \$2,000(P/A,15%,10) - \$10,000 = \$2,000(5.019) - \$10,000 = +\$38.
Project is acceptable using this criterion.

(e) The solutions in (b) and (d) represent only the proposed investment. By the choice of a MARR of 15%, the company presupposes that any reinvested cash throw-offs (from whatever source) will earn 15% or more. However, the method in (a) restricts a portion of the throw-offs to earning only 6% throughout the life of the project. The method assumes a compound investment of actually separable parts and thus the method itself contains a conceptual error. This error in concept is not overcome simply by using the MARR as in part (c) simply because the correct decision could be made. The disposition of future cash receipts is a future problem; a long range capital budgeting problem.

C-10

(a) F = \$2,000(F/A,6%,10)=\$2,000(13.181) = \$26,362
$(F/P,i_s\%,10)$ = \$26,362/\$10,000 = 2.6362 from which
i_s = 10.2%.

(b) From Problem C-9 (b), i = 15.1%; the project is acceptable.

(c) F = \$2,000(F/A,15%,10) = \$2,000(20.304) = \$40,608.
$(F/P,i_s\%,10)$ = \$40,608/\$10,000 = 4.0608 from which
i_s = 15.0%.

(d) From Problem C-9 (d), NPW = \$38; the project is acceptable.

(e) See part (e) of Problem C-9. This method understates the true rate of return even more than the Hoskold-type method because all, rather than just a portion, of the cash flow throw-offs are assumed to be reinvested.

APPENDIX F

Notes on Certain Changes in the Income Tax Laws of the United States

F-1

(a) 0 = -$180,000 + $46,800(P/A,i%,8)
i = approximately 19.9%

(b) Investment tax credit at year zero = 0.10($180,000) = $18,000
Annual income tax = 0.46($46,800-$180,000/8) = $11,178
0 = -$180,000 + $18,000 + ($46,800-$11,178)(P/A,i%,8)
i = approximately 14.6%

(c) Years-digits depreciation = 8/36($180,000) = $40,000 for year 1 and will decrease by $5,000 each year. Income taxes are 0.46($46,800-$40,000) = $3,128 for year 1 and will increase by 0.46($5,000) = $2,300 each year.
0 = -$180,000 + $18,000 + ($46,800-$3,128)(P/A,i%,8)
- $2,300(P/G,i%,8)
i = approximately 16.1%

(d) Depreciation for years 1 through 5 is $180,000/5 = $36,000
Income tax for years 1 through 5 is 0.46($46,800-$36,000)
= $4,968
Income tax for years 6 through 8 is 0.46($46,800) = $21,528
0 = -$180,000 + $18,000 + ($46,800-$4,968)(P/A,i%,5)
+ ($46,800-$21,528)(P/A,i%,8 - P/A,i%,5)
i = approximately 16.4%

F-2

Year	Cash Flow Before Income Taxes	Write-off of Initial Outlay for tax Purposes	Influence on Taxable Income	Influence of Income Taxes on Cash Flow -0.46C	Cash Flow After Income Taxes A + D
	A	B	C	D	E
0	-$180,000			+$18,000	-$162,000
1	+46,800	-$ 27,000	+$19,800	-9,108	+37,692
2	+46,800	-39,600	+7,200	-3,312	+43,488
3	+46,800	-37,800	+9,000	-4,140	+42,660
4	+46,800	-37,800	+9,000	-4,140	+42,660
5	+46,800	-37,800	+9,000	-4,140	+42,660
6	+46,800	0	+46,800	-21,528	+25,272
7	+46,800	0	+46,800	-21,528	+25,272
8	+46,800	0	+46,800	-21,528	+25,272
	+$194,400	-$180,000	+194,400	-$71,424	+$122,976

$$0 = -\$162{,}000 + \$37{,}692(P/F,i\%,1) + \$43{,}488(P/F,i\%,2)$$
$$+\$42{,}660(P/A,i\%,5-P/A,i\%,2) + \$25{,}272(P/A,i\%,8-P/A,i\%,5)$$

i = approximately 16.2%

F-3

Use factors from Table D-30 as applied to the cash flow for each year that was calculated in Column E of the foregoing solution to Problem F-2. As explained on page 584, the factor for the year -1 to 0 is the reciprocal of the factor for year 0 to 1.

Year	Cash flow	15% PW factor	PW @ 15%	20% PW factor	PW @ 20%
-1 to 0	-$162,000	1.0715	-$173,583	1.0940	-$177,228
0 to 1	+37,692	0.9333	+35,178	0.9141	+34,454
1 to 2	+43,488	0.8115	+35,291	0.7618	+33,129
2 to 3	+42,660	0.7057	+30,105	0.6348	+27,081
3 to 4	+42,660	0.6136	+26,176	0.5290	+22,567
4 to 5	+42,660	0.5336	+22,763	0.4408	+18,805
5 to 6	+25,272	0.4640	+11,726	0.3674	+ 9,285
6 to 7	+25,272	0.4035	+10,197	0.3061	+ 7,736
7 to 8	+25,272	0.3508	+ 8,865	0.2551	+ 6,447
			+$6,719		-$17,724

$$i = 15\% + \frac{6{,}719}{24{,}443}(5\%) = \text{approximately } 16.4\%$$

F-4

Headings are the same as in the solution to Problem F-2.

Year	A	B	C	D	E
0	-$180,000			+$18,000	-$162,000
1	+46,800	-$32,400	+$14,400	-6,624	+40,176
2	+46,800	-59,400	-12,600	+5,796	+52,596
3	+46,800	-45,000	+ 1,800	-828	+45,972
4	+46,800	-28,800	+18,000	-8,280	+38,520
5	+46,800	-14,400	+32,400	-14,904	+31,896
6	+46,800	0	+46,800	-21,528	+25,272
7	+46,800	0	+46,800	-21,528	+25,272
8	+46,800	0	+46,800	-21,528	+25,272
	+$194,400	-$180,000	+$194,400	-$71,424	+$122,976

0 = -$162,000 + $40,176(P/F,i%,1) + $52,596(P/F,i%,2)
+$45,972(P/F,i%,3) + $38,520(P/F,i%,4) + $31,896(P/F,i%,5)
+$25,272(P/A,i%,8 - P/A,i%,5)

i = approximately 17.0%

F-5

Headings are the same as in the solution to Problem F-2.

The only differences from the table in Problem F-4 are in the figures for columns B, C, D, and E for years 1-3, as follows:

Year	B	C	D	E
1	-$36,000	+$10,800	-$4,968	+$41,832
2	-57,600	-10,800	+4,968	+51,768
3	-43,200	+3,600	-1,656	+45,144

0 = -$162,000 + $41,832(P/F,i%,1) + $51,768(P/F,i%,2)
+$45,144(P/F,i%,3) + $38,520(P/F,i%,4) + $31,890(P/F,i%,5)
+ $25,272(P/A,i%,8 - P/A,i%,5)

i = approximately 17.1%

F-6

The following table that applies the 1985 percentages uses the same headings as the solution to Problem F-2.

Year	A	B	C	D	E
0	-$110,000			+$11,000	-$99,000
1	+26,300	-$ 19,800	+$ 6,500	-3,185	+23,115
2	+26,300	-36,300	-10,000	+4,900	+31,200
3	+26,300	-27,500	-1,200	+588	+26,888
4	+26,300	-17,600	+8,700	-4,263	+22,037
5	+26,300	-8,800	+17,500	-8,575	+17,725
6-10	+26,300 per year	0 per year	+26,300 per year	-12,887 per year	+13,413 per year
	+$153,000	-$110,000	+$153,000	-$63,970	+$89,030

0 = -$99,000 + $23,115(P/F,i%,1) + $31,200(P/F,i%,2) +$26,888(P/F,i%,3) + $22,037(P/F,i%,4) + $17,725(P/F,i%,5) + $13,413(P/A,i%,10 - P/A,i%,5)

i = approximately 16.9%

The 1986 percentages change the figures for columns B, C, D, and E for years 1, 2, and 3 as follows:

	B	C	D	E
1	-$22,000	+$4,300	-$2,107	+$24,193
2	-35,200	- 8,900	+4,301	+30,661
3	-26,400	-100	+49	+26,349

0 = -$99,000 + $24,193(P/F,i%,1) + $30,661(P/F,i%,2) +$26,349(P/F,i%,3) + $22,037(P/F,i%,4) + $17,725(P/F,i%,5) + $13,413(P/A,i%,10 - P/A,i%,5)

i = approximately 17.0%

F-7

(a) 0 = -$165,000 + $36,850(P/A,i%,10)
i = approximately 18.1%

(b) Investment tax credit at year zero = 0.10($165,000) = $16,500
Annual income tax = 0.46($36,850-$165,000/10) = $9,361

0 = -$165,000 + $16,500 + ($36,850-$9,361)(P/A,i%,10)

i = approximately 13.1%

F-7 continued

(c) Years-digits depreciation = 10/55($165,000) = $30,000 for year 1 and will decrease by $3,000 each year. Income tax is 0.46($36,850 - $30,000) = $3,151 for year 1 and will increase by 0.46($3,000) = $1,380 each year.

0 = -$165,000 + $16,500 + ($36,850 - $3,151)(P/A,i%,10) - $1,380(P/G,i%,10)

i = approximately 14.5%

(d) Depreciation for years 1 through 5 is $165,000/5 = $33,000
Income tax for years 1 through 5 is 0.46($36,850-$33,000) = $1,771
Income tax for years 6 through 10 is 0.46($36,850) = $16,951

0 = -$165,000 + $16,500 + ($36,850 - $1,771)(P/A,i%,5) + ($36,850 - $16,951)(P/A,i%,10 - P/A,i%,5)

i = approximately 15.5%

F-8

Headings are the same as in the solution to Problem F-2.

Year	A	B	C	D	E
0	-$165,000			+$16,500	-$148,500
1	+36,850	-$24,750	+$12,100	-5,566	+31,284
2	+36,850	-36,300	+550	-253	+36,597
3	+36,850	-34,650	+2,200	-1,012	+35,838
4	+36,850	-34,650	+2,200	-1,012	+35,838
5	+36,850	-34,650	+2,200	-1,012	+35,838
6-10	+36,850 per year	0 per year	+36,850 per year	-16,951 per year	+19,899 per year
	+$203,500	-$165,000	+$203,500	-$77,110	+$126,390

0 = -$148,500 +$31,284(P/F,i%,1) + $36,597(P/F,i%,2) + $35,838(P/A,i%,5 - P/A,i%,2) + $19,899(P/A,i%,10 - P/A,i%,5)

i = approximately 15.3%

F-9

(a) $0 = -\$180{,}000 + \$32{,}000(P/A,i\%,15)$

i = approximately 15.8%

(b) Investment tax credit at year zero = 0.10($180,000) = $18,000
Annual income tax = 0.46($32,000 - $180,000/15) = $9,200

$0 = -\$180{,}000 + \$18{,}000 + (\$32{,}000 - \$9{,}200)(P/A,i\%,15)$

i = approximately 11.2%

(c) Years-digits depreciation = $\frac{15}{120}(\$180{,}000)$ = $22,500 for year 1 and will decrease by $1,500 each year. Income tax is 0.46($32,000-$22,500) = $4,370 for year 1 and will increase by 0.46($1,500) = $690 each year.

$$0 = -\$180{,}000 + \$18{,}000 + (\$32{,}000 - \$4{,}370)(P/A,i\%,15) - \$690(P/G,i\%,15)$$

i = approximately 12.4%

(d) Depreciation for years 1 through 5 is $180,000/5 = $36,000
Because the depreciation deduction of $36,000 is $4,000 greater than the before-tax taxable income of $32,000, income taxes will be reduced by 0.46($4,000) = $1,840 a year for the first 5 years. After-tax cash flow for years 1 through 5 will be $32,000 + $1,840 = $33,840. Income tax each year for years 6 through 15 is 0.46($32,000) = $14,720.

$$0 = -\$180{,}000 + \$18{,}000 + \$33{,}840(P/A,i\%,5) + (\$32{,}000 - \$14{,}720)(P/A,i\%,15 - P/A,i\%,5)$$

i = approximately 14.2%

F-10

Headings are the same as in the solution to Problem F-2.

Year	A	B	C	D	E
0	-$180,000			+$18,000	-$162,000
1	+32,000	-$27,000	+$5,000	-2,300	+29,700
2	+32,000	-39,600	-7,600	+3,496	+35,496
3	+32,000	-37,800	-5,800	+2,668	+34,668
4	+32,000	-37,800	-5,800	+2,668	+34,668
5	+32,000	-37,800	-5,800	+2,668	+34,668
6-15	+32,000 per year	0 per year	+32,000 per year	-14,720 per year	+17,200 per year
	+$300,000	-$180,000	+$300,000	-$120,000	+$180,000

$$0 = -\$162{,}000 + \$29{,}700(P/F,i\%,1) + \$35{,}496(P/F,i\%,2) + \$34{,}668(P/A,i\%,5 - P/A,i\%,2) + \$17{,}280(P/A,i\%,15 - P/A,i\%,5)$$

i = approximately 14.0%

F-11

This is similar to the solution of Problem F-3

Year	Cash flow	12% PW factor	PW @ 12%	15% PW factor	PW @ 15%
-1 to 0	-$162,000	1.0578	-$171,364	1.0715	-$173,583
0 to 1	+29,700	0.9454	+28,078	0.9333	+27,719
1 to 2	+35,496	0.8441	+29,962	0.8115	+28,805
2 to 3 through 4 to 5	+34,668 per year	2.027	+70,272	1.853	+64,240
5 to 6 through 14 to 15	+17,280 per year	3.395	+58,666	2.678	$46,276
			+$15,614		-$6,543

$$i = 12\% + \frac{15,614}{22,157}(3\%) = \text{approximately } 14.1\%$$

F-12

Headings are the same as in the solution to Problem F-2.

Year	A	B	C	D	E
0	-$180,000			+$18,000	-$162,000
1	+32,000	-$36,000	-$4,000	+1,840	+33,840
2	+32,000	-57,600	-25,600	+11,776	+43,776
3	+32,000	-43,200	-11,200	+5,152	+37,152
4	+32,000	-28,800	+3,200	-1,472	+30,528
5	+32,000	-14,400	+17,600	-8,096	+23,904
6-15	+32,000 per year	0 per year	+32,000 per year	-14,720 per year	+17,280 per year
	+$300,000	-$180,000	+$300,000	-$120,000	+$180,000

0 = -$162,000 + $33,840(P/F,i%,1) + $43,776(P/F,i%,2)
+$37,152(P/F,i%,3) + $30,528(P/F,i%,4) + $23,904(P/F,i%,5)
+$17,280(P/A,i%,15 - P/A,i%,5)

i = approximately 14.6%

F-13

Headings correspond to those in the solution to Problem F-2 but the tax (column D) for years 1-15 is now 0.514C. The investment tax credit of $18,000 is unchanged from the solution to Problem F-10.

Year	A	B	C	D	E
0	-$180,000			+$18,000	-$162,000
1	+32,000	-$27,000	-$5,000	-2,570	+29,430
2	+32,000	-39,600	-7,600	+3,906.4	+35,906.4
3	+32,000	-37,800	-5,800	+2,981.2	+34,981.2
4	+32,000	-37,800	-5,800	+2,981.2	+34,981.2
5	+32,000	-37,800	-5,800	+2,981.2	+34,981.2
6-15	+32,000 per year	0 per year	+32,000 per year	-16,448 per year	+15,552 per year
	+$300,000	-$180,000	+$300,000	-$136,200	+$163,800

$$0 = -\$162{,}000 + \$29{,}430(P/F,i\%,1) + \$35{,}906.4(P/F,i\%,2) + \$34{,}981.2(P/A,i\%,5-P/A,i\%,2) + \$15{,}552(P/A,i\%,15 - P/A,i\%,5)$$

i = approximately 13.4%

F-14

The state income tax first needs to be computed. The headings are the same as in Problem F-13 except that the tax disbursement D is now 0.10C.

Year	A	B	C	D	E
0	-$180,000				-$180,000
1-15	+32,000 per year	-$12,000 per year	+$20,000 per year	-$2,000 per year	+30,000 per year
	+$300,000	-$180,000	+$300,000	-$30,000	+$270,000

The net positive before-tax cash flow of $32,000 in Column A presumably was derived from combining estimated positive and negative cash flows. The analysis assumes that each dollar of positive or negative cash flow changes the year's taxable income by one dollar. For the federal analysis, each year's state income tax of $2,000 is a negative cash flow that reduces the federal before-tax cash flow and the federal taxable income by $2,000. The figures from column E in the state analysis may therefore be used for column A in the federal analysis, which assumes a tax rate of 46%.

F-14 continued

Year	A	B	C	D	E
0	-$180,000			+$18,000	-$162,000
1	+30,000	-$27,000	+$3,000	-1,380	+28,620
2	+30,000	-39,600	-9,600	+4,416	+34,416
3	+30,000	-37,800	-7,800	+3,588	+33,588
4	+30,000	-37,800	-7,800	+3,588	+33,588
5	+30,000	-37,800	-7,800	+3,588	+33,588
6-15	+30,000 per year	0 per year	+30,000 per year	-13,800 per year	+16,200 per year
	+$270,000	-$180,000	+$270,000	-$106,200	+$163,800

0 = -$162,000 + $28,620(P/F,i%,1) + $34,416(P/F,i%,2) +$33,588(P/A,i%,5 - P/A,i%,2) + $16,200(P/A,i%,15 - P/A,i%,5)

i = approximately 13.0%

Note: a check on any such calculation of after-tax cash flow is to note that the net after-tax cash flow (+$163,800 in Problems F-13 and F-14) will be the same regardless of the timing of the depreciation deductions as long as the same applicable tax rate is assumed throughout the entire study period.

F-15

(a) With 100% immediate write-off, the prospective after-tax rate of return will be the same as the prospective before-tax rate if a constant applicable tax rate is assumed throughout the entire study period. To find before-tax rate of return in this case, solve:

0 = -$5,000 + $1,100(P/A,i%,10)

i = approximately 17.7%

The immediate write-off will reduce the net investment by 15% and will reduce the annual positive before-tax cash flow by 15% (because no depreciation deduction will be available).

0 = -$5,000 + $750 + ($1,100 - $165)(P/A,i%,10)

Of course i continues to be approximately 17.7%

F-15 continued

(b) Headings are the same as in the solution to Problem F-2 except that Column D is 0.15C for years 1-10.

Year	A	B	C	D	E
0	-$5,000			+$500	-$4,500
1	+1,100	-$750	+$350	-52.5	+1,047.5
2	+1,100	-1,100	0	0	+1,100
3	+1,100	-1,050	+50	-7.5	+1,092.5
4	+1,100	-1,050	+50	-7.5	+1,092.5
5	+1,100	-1,050	+50	-7.5	+1,092.5
6-10	+1,100 per year	0 per year	+1,100 per year	-165 per year	+935 per year
	+$6,000	-$5,000	+$6,000	-$400	+$5,600

0 = -$4,500 + $1,047.5(P/F,i%,1) + $1,100(P/F,i%,2)
+ $1,092.5(P/A,i%,5 - P/A,i%,2) + $935(P/A,i%,10 - P/A,i%,5)

i = approximately 19.1%

F-16

In year zero (a) reduces taxes by the difference between 15% of $5,000 and 10% of $5,000; this is $250. In year 1 (a) increases taxes by 15% of 0.15($5,000) = $112.5; similar calculations will yield the stated figures for years 2, 3, 4, and 5.

0 = +$250 - $112.5(P/F,i%,1) - $165(P/F,i%,2)
-$157.5(P/A,i%,5 - P/A,i%,2)

A solution of the foregoing equation shows that the cost of the money made available by selecting (a) is approximately 49% after income taxes.

F-17

The cash flow difference at zero date is 0.30($5,000) - 0.10($5,000) = +$1,000. Cash flow differences at dates 1 through 5 are 30% of the allowable depreciation charges for each year. These differences are, respectively, -$225, -$330, -$315, -$315, and -$315.

0 = +$1,000 - $225(P/F,i%,1) - $330(P/F,i%,2)
- $315(P/A,i%,5 - P/A,i%,2)

F-17 continued

A solution of the foregoing equation shows that the cost of the money made available by selecting (a) is approximately 14.6% after income taxes.

F-18

An analysis similar to that in Problem F-17 yields the following equation:

$$0 = + \$1,800 - \$345(P/F,i\%,1) - \$506(P/F,i\%,2) - \$483(P/A,i\%,5 - P/A,i\%,2)$$

The after-tax cost of this money is now 8.4%

F-19

A higher incremental tax rate in year 1 or thereafter does not influence the difference in cash flow at year zero. Therefore the prospect of a higher incremental tax rate increases the after-tax cost of the money made available by selecting the 100% write-off. It follows that the attractiveness of the 100% write-off is decreased by the expectation that the applicable tax rate will increase in, say, the next five years. (This is a common expectation for successful small businesses, both incorporated and unincorporated.)

F-20

This is a more complex question than it seems to be at first glance. The conclusions from a numerical example will be influenced by the assumptions made about the depreciation percentages allowable in years to come with the 20% first-year depreciation on $20,000 and with the 100% immediate write-off of $5,000. The assumption about the exact timing of the first-year depreciation is also relevant. A possible solution is the following:

Consider a total investment of $20,000 in 1982. Let (a) represent the 1982 rule and (b) the 1980 rule. Assume that the investment is in a 5-year property as defined in the tax act of 1981. In (a) assume that $5,000 of this investment will be chosen for the 100% write-off. Contrast this with (b) in which there will be additional first-year depreciation of $4,000 (20% of $20,000). In (a) there will be an investment tax credit of $1,500 (10% of the $15,000 not subject to the 100% write-off). In (b) the investment tax credit will be $2,000 (10% of the entire investment). Depreciation deductions will be as follows:

F-20 continued

Year	(a)	(b)	(a) - (b)
0	$5,000	$ 0	$5,000
1	2,250	6,400	-4,150
2	3,300	3,520	-220
3	3,150	3,360	-210
4	3,150	3,360	-210
5	3,150	3,360	-210
	$20,000	$20,000	0

At date zero, the $5,000 immediate write-off will reduce zero-year taxes by 0.40($5,000) = $2,000. But this will cause the loss of $500 of investment tax credit, so the net positive cash flow from (a) as compared to (b) is $2,000 - $500 = $1,500. This will be offset in years 1-5 by negative cash flow of 40% of the extra depreciation allowable with (b). Therefore, the relevant cash flow series to be evaluated is 0, +$1,500; 1, -$1,660; 2, -$88; 3, -$84; 4, -$84; 5, -$84. This series has a present worth of zero at an i of approximately 24%. With these assumptions, the 1982 rule would not be superior to the 1980 rule unless our sole proprietor would find it advantageous to finance by borrowing money at an after-tax cost of 24%.

F-21

The bracket creep occurs particularly after a corporation's taxable income reaches $50,000. Its effect may be illustrated by assuming a corporation with a taxable income of $50,000 in 1981. Assume that a 10% inflation each year is accompanied by a 10% increase in taxable income; in other words, the company's before-tax income just keeps up with inflation.

Year	Taxable income	Tax	Percentage increase in tax compared to preceding year
1981	$50,000	$9,250	
1982	55,000	10,250	9.8%
1983	60,500	11,400	11.2%
1984	66,550	13,215	15.9%
1985	73,205	15,211	15.1%

Similar results will be obtained starting with figures higher than $50,000.

F-22

Sam's 1980 tax (from Table 12-8) was \$2,605. Jennifer's was \$7,434 + 0.44(\$30,000 - \$29,800) = \$7,522. Under the 1980 schedule an income of \$17,600 would have had a tax of \$2,605 + 0.30(\$17,600 - \$16,000) = \$3,085. Sam's 1981 tax is \$3,085 - 0.0125(\$3,085) = \$3,046.44. This is an increase of 16.9% above his 1980 tax.

Under the 1980 schedule an income of \$33,000 would have had a tax of \$7,434 + 0.44(\$33,000 - \$29,800) = \$8,842. Jennifer's 1981 tax is \$8,842 - 0.0125(\$8,842) = \$8,731.47. This is an increase of 16.1% above her 1980 tax.